BARRON'S

AP®

PHYSICS 1
PREMIUM
WITH 4 PRACTICE TESTS

SECOND EDITION

Kenneth Rideout, M.S.
Physics Teacher
Science Department Chairperson
Wayland High School
Wayland, Massachusetts

Jonathan S. Wolf, M.A., Ed.M.
Adjunct Professor of Physics
Fairleigh Dickinson University
Madison, New Jersey

About the Authors:

Ken Rideout has a B.S. in Honors Physics from Purdue University and an M.S. in Physics from Carnegie Mellon University. He has been teaching high school physics for more than fifteen years in the Boston area and is currently the science department head at Wayland High School in Wayland, MA.

Jonathan Wolf has been teaching physics at both the secondary school and college levels for more than thirty-five years, and has also been an adjunct assistant professor of astronomy at Hofstra University. He has published over forty professional papers in the fields of astronomy, physics, and physics education and served for more than ten years as assistant editor for *The Science Teachers Bulletin* published by the Science Teachers Association of New York State (STANYS). In addition to being the author of Barron's *AP Physics B*, he is also the author of Barron's *College Review: Physics* and previously was co-author of Barron's *SAT Subject Test in Physics*. Jonathan Wolf is currently an adjunct professor of physics at Fairleigh Dickinson University.

Published by Kaplan, Inc., d/b/a Barron's Educational Series
750 Third Avenue
New York, NY 10017
www.barronseduc.com

ISBN: 978-1-5062-6210-9

10 9 8 7 6 5 4 3 2 1

Kaplan, Inc., d/b/a Barron's Educational Series print books are available at special quantity discounts to use for sales promotions, employee premiums, or educational purposes. For more information or to purchase books, please call the Simon & Schuster special sales department at 866-506-1949.

Contents

PRACTICE TESTS

APPENDIX

As you review the content in this book to work toward earning that **5** on your AP Physics 1 exam, here are five things that you **MUST** know above everything else.

Barron's Essential (AP Physics 1)

1 **Know your kinematics:**

- Know the difference between velocity and speed, displacement and distance.
- Know to use equations of motion only for problems involving constant acceleration and to otherwise use graphical methods to analyze the kinematics.
- Find hidden information within the problem (initial or final speeds of zero, accelerations of 9.8 m/s^2).
- Keep your *x*- and *y*-motions separate: the only connecting variable is time.

2 **Know your dynamics:**

- Start all analyses with a free-body diagram.
- Align your coordinate system with the direction of acceleration (if known).
- Be alert for situations that require a net force but no corresponding change in speed (centripetal forces).
- Do not add additional forces in an ad hoc manner. Decide on the number of forces based on the object's interactions, not its motion.

3 **Know your conservation laws:**

- If all the masses and motions of interacting particles are specified, conserve the net values of energy, momentum, and angular momentum.
- If the object in question is interacting with an object whose masses and motions are unspecified, then use the interactions to calculate changes in "conserved" quantities.
- When using conservation of energy, be alert to work done by nonconservative forces.
- When using conservation of momenta, remember that both linear and angular momenta are vector quantities.

4 **Look for crosscutting questions:**

- This exam is designed to probe for your understanding of the connections between the various topics covered in a first year physics class. It is not enough to be able to solve the "classic" problem types. You must be able to solve multitiered problems (e.g., apply conservation of angular momentum and dynamics at the same time within one problem) and apply your knowledge to new situations.

5 **Understand the underlying concepts:**

- Be sure to understand how to explain the why's, not just to calculate a numerical solution. Novel situations or new types of problems can be solved by students who understand the concepts behind the equations.

Preface

In this review book, you will find all the material you need to review and prepare for the AP Physics 1 exam, a first-year precalculus course. This book is a newly updated, targeted, and enriched version of Barron's *AP Physics 1*.

Each chapter has review questions that vary in style and level of difficulty. These are intended to test your level of understanding of the review material. Some of these chapter questions may be easier, harder, or in a slightly different style from the actual AP exam as they are limited to the content of that particular chapter. Likewise, each chapter has a mix of traditional problem-solving examples as well as the more text-based and conceptual type of questions that appear on the AP Physics 1 exam. The tests, on the other hand, are broad in scope and draw from several different content areas at once. Included in this book, as well as online, are several full-length practice tests that mirror the actual AP exams in style, content, and difficulty. All questions have full solutions and explanations provided. Before the review chapters is a diagnostic test that can be used for practice or to guide you specifically to certain content areas you may need to review before the AP exam. Additional problem-solving strategies are provided throughout the book.

As the College Board has better defined the scope of the relatively new AP Physics 1 exam, we are better able to target this book for students who are preparing for that exam. This material, originally from a previous book written by Jonathan Wolf, has been updated and has benefited from several expert reviewers and editors at Barron's. I would like to thank my many students through the years who have made my job as a physics teacher a real joy. Also, I especially want to thank my wife, Irene, for her support throughout this project.

Ken Rideout
Wayland, Massachusetts
December 2019

Introduction

STRUCTURE AND SCOPE OF THE AP PHYSICS EXAMS

The College Board currently offers four AP Physics exams. This book is for students preparing for the AP Physics 1 exam, which corresponds to a first-year algebra-based college course. The other exams are AP Physics 2 (a second-year algebra-based college course), AP Physics C Mechanics (calculus based), and AP Physics C Electricity and Magnetism (also calculus based). Note that the old AP Physics B exam is a retired test and is no longer offered by the College Board. Both the AP Physics 1 and AP Physics 2 exams focus on conceptual under-pinnings and basic scientific reasoning along with the traditional problem-solving aspects of physics. In addition, both exams have questions that require experiential lab understanding. Although there are some calculation-oriented questions, these two tests

> **AP Physics 1 Test Basics**
> - 90 minutes for 50 multiple-choice questions
> - 90 minutes for 5 free-response questions
> - Calculator allowed throughout
> - Formula sheet and table of information provided throughout

are explicitly designed not to be "plug and chug" type questions. Students who do not thoroughly understand the physics concepts behind the equations will find themselves at a disadvantage.

The AP Physics 1 exam focuses on mechanics (including rotational mechanics), the three con-servation laws (energy, linear momentum, and angular momentum), electric charge and force, introductory circuits, and mechanical waves (including sound). The AP Physics 2 exam picks up many of the other topics typically found in an introductory sequence in college physics and also covers some of the overlapping areas, such as force fields, capacitors, and electromagnetic waves. Formulas are provided for you during the test (see the appendix). However, it is important that you not only know where and what the provided equations mean, but you must also be able to determine quickly under what situations the equations can and should be used. Even if a question is conceptual, having a corresponding equation in mind can guide your thinking. Note that some useful equations are not given (e.g., orbital velocity, Kepler's third law, moment of inertia, projectile motion) and that some given equations (universal gravitational potential energy, for example) are unlikely to be used. The equation sheet does, however, provide a solid foundation. A well-prepared student will be able to find and understand every equation on it.

Although you may not use your calculator often during the exam, one is allowed throughout. Check the College Board website for an approved list of calculators. Generally, all scientific and graphing calculators are allowed. You should make sure you have extra batteries for your calculator during the exam. A ruler is also permitted, but its usefulness is likely limited to drawing straight lines, if needed, during the free-response section.

ORGANIZATION OF THIS BOOK

This introduction serves as general background to the AP exam. A diagnostic test follows. This exam is intended as a tool to help already-prepared students determine if they have any weakness in content and to direct them to the appropriate chapter in this book for review. Although example problems and targeted questions are included at the end of each chapter to reinforce the content, the full-length practice tests are where you will find questions that are most closely modeled after the actual AP Physics 1 exam. Each chapter covers a specific topic in physics, so the questions in each chapter are limited to that specific concept. The specific problem types in the AP Physics 1 exam are mirrored in the practice tests. For example, "choose two" multiple-choice problems, experimental design questions, paragraph-long explanations, and text-heavy conceptual questions are all included.

UNITS

Preparing for an AP exam takes time and planning. In fact, your preparation should begin in September when you start the class. If you are using this review book during the year, the content review chapters should parallel what you are covering in class. If you are using this review book a few weeks prior to the exam in May, your strategy needs to change. The review material should help you refresh your memory as you work on the practice tests. In either case, you should have a plan.

In this chapter, we will look at study skills and tips for helping you do well on the AP Physics 1 exam. One of the most important things to remember is that most physical quantities have units associated with them. You must memorize units since you may be asked questions about them in the multiple-choice section. In the free-response questions, you must include all units when using equations, making substitutions, and writing final answers.

A list of standard fundamental (SI) units as well as a list of some derived units are shown in the following two tables. As you work through the different chapters, make a note (for example, on index cards) of each unit.

Make sure you set up a review schedule.

Make sure you memorize all units. Be sure to include them with all calculations and final answers.

Table 1. Fundamental SI Units Used in Physics

Quantity	Unit Name	Abbreviation
Length	Meter	m
Mass	Kilogram	kg
Time	Second	s
Electric current	Ampere	A
Temperature	Kelvin	K
Amount of substance	Mole	mol
Luminous intensity	Candela	cd*

*Not likely to appear on the AP Physics 1 exam

Table 2. Some Derived SI Units Used in Physics

Quantity	Unit Name	Abbreviation	Expression in Other SI Units
Area			m^2
Linear velocity			m/s
Linear acceleration			m/s^2
Force	Newton	N	$kg \cdot m/s^2$
Momentum			$kg \cdot m/s$
Impulse			$N \cdot s = kg \cdot m/s$
Angular velocity			rad/s
Angular acceleration			rad/s^2
Torque			Nm
Angular momentum			$kg \cdot m^2/s$
Moment of inertia			$kg \cdot m^2$
Spring constant		N/m	kg/s^2
Frequency	Hertz	Hz	s^{-1}
Pressure	Pascal	Pa	$N/m^2 = kg/(m \cdot s^2)$
Work, energy	Joule	J	$N \cdot m = kg \cdot m^2/s^2$
Power	Watt	W	$J/s = kg \cdot m^2/s^3$
Electric charge	Coulomb	C	$A \cdot s$
Electric potential	Volt	V	$J/C = kg \cdot m^2/(A \cdot s^3)$
Resistance	Ohm	Ω	$V/A = kg \cdot m^2/(A^2 \cdot s^3)$

RELATIONSHIPS AND REVIEW OF MATHEMATICS

Since AP Physics 1 is an algebra-based course, the appendix reviews some essential aspects of algebra. In physics, we often discuss how quantities vary using proportional relationships. Four special relationships are commonly used. You can review them in more detail by referring to the appendix. You should memorize these relationships.

> **REMINDER**
>
> These relationships are also useful for analyzing data to answer laboratory-based questions. A laboratory-based question is always on the exam.

- **Direct relationship**—This is usually represented by the algebraic formula $y = kx$, where k is a constant. This is the equation of a straight line, starting from the origin. An example of this relationship is Newton's second law of motion, $\vec{a} = \dfrac{\vec{F}_{net}}{m}$, which states that the acceleration of a body is directly proportional to the net force applied (see Chapter 3).

- **Inverse relationship**—This is usually represented by the algebraic formula $y = \dfrac{k}{x}$. This is the equation of a hyperbola. An example of this relationship can be seen in a different version of Newton's second law, $\vec{F}_{net} = m\vec{a}$. In this version, if a constant net force is applied to a body, the mass and acceleration are inversely proportional to each other. Some special relationships, such as gravitation and static electrical forces, are known as inverse square law relationships. The forces are inversely proportional to the square of the distances between the two bodies (see Chapters 5 and 9).

- **Squared (quadratic) relationship**—This is usually represented by the algebraic formula $y = kx^2$ and is the equation of a parabola starting from the origin. An example of this relationship can be seen in the relationship between the displacement and uniform acceleration of a mass from rest $\vec{d} = \frac{1}{2}\vec{a}t^2$ (see Chapter 2).

- **Square root relationship**—This is usually represented by the algebraic formula $y = k\sqrt{x}$ and is the equation of a "sideways" parabola. This relationship can be seen in the relationship between the period of a simple pendulum and its length, $T = 2\pi\sqrt{L/g}$ (see Chapter 8).

- As you review your material, you should know each of these relationships and their associated graphs (see the appendix for more details).

TIPS FOR ANSWERING MULTIPLE-CHOICE QUESTIONS

Without a doubt, multiple-choice questions can be tricky. The AP Physics 1 exam contains 50 multiple-choice questions. These can range from a simple recall of information to questions about units, graphs, proportional relationships, formula manipulations, and simple calculations. According to the College Board, the breakdown of topics covered on the multiple-choice section is as follows:

Kinematics:	10–16%
Forces (Dynamics):	12–18%
Energy:	16–24%
Gravity & Circular Motion:	4–6%
Linear Momentum:	10–16%
Rotational Motion:	10–16%
Oscillations (Simple Harmonic Motion):	2–4%
Electric Charge & Force:	4–6%
Simple Circuits:	6–8%
Waves:	12–16%

One tip to remember is that there is no penalty for wrong answers. This means that you want to try to answer all questions. Instead of randomly guessing, however, you can improve your chances of getting a correct answer if you can eliminate at least two answer choices. Guess intelligently. For the Physics 1 exam, all multiple-choice questions will have four answer choices. A new question type has also been added, the "multiple-correct items" question. These questions, which ask you to mark two correct responses, will be clearly indicated on the exam and are introduced in this book. These "choose two" question types are scored all-or-nothing, meaning you must get both correct to get credit. There will be five of these "choose two" questions at the end of the multiple-choice section (45 single-select questions followed by this set of 5 choose-two questions).

When you read a multiple-choice question, try to get to the essential aspects. You have 90 minutes for this part, so do not waste too much time per question. Try to eliminate two or three choices. If a formula is needed, you may try to use approximations (or simple multiplication and division). For example, the magnitude of the acceleration due to gravity (g) can be approximated as 10 m/s^2. You can also use estimations or order of magnitude approximations to see if answers make sense.

As you work on the multiple-choice questions in the practice tests, look for distractors. These are choices that may look reasonable but are incorrect. For example, if the question is expecting you to divide to get an answer, the distractor may be an answer obtained by multiplying. Watch out for quadratics (such as centripetal force) or inverse squares (such as gravitation), where linear reasoning does not apply.

If you cannot recall some information, perhaps another similar question will cue you as to what you need to know. (You may work on only one part of the exam at a time.) When you read the question, try to link it to the overall general topic, such as kinematics, dynamics, or electricity, and then narrow down the specific area and the associated formula. Finally, you must know which quantities are vectors and which quantities are scalars (see Chapter 1).

Each multiple-choice question in the practice tests is cross-indexed with the general topic area of physics to guide you on your review. As you work on the tests and check your answers, you can easily go back to the topic area to review. At the start of your review, you may want to work on the multiple-choice questions untimed for the diagnostic and first practice tests. A few days before the AP exam (see the timeline schedule later in this chapter), you should do the last practice test timed (90 minutes for the multiple-choice section and an additional 90 minutes for the free-response section).

TIPS FOR SOLVING FREE-RESPONSE QUESTIONS

The AP Physics 1 exam includes five free-response questions:

   1 experimental design question (12 points)

   1 qualitative/quantitative translation problem (12 points)

   1 paragraph style short answer (7 points)

   2 short-answer questions (7 points each)

You will have 90 minutes for this section. Each question will have specific point allocations for each subsection. However, the free-response portion of the exam in its entirety is worth 50 percent of your "raw score." Keep in mind that the curve determining your actual reported score (out of 5), based on the raw score from the points you earned, changes from year to year.

Qualitative/quantitative translation problems emphasize proportional and symbolic reasoning skills. These problems also provide an opportunity for you to demonstrate your ability to translate between multiple representations of the same problem. Every exam has questions that require students to "justify your response." There is an expectation that you will be able to write paragraph-long coherent answers involving multiple concepts in physics. You are expected to be able to read critically and write precise and coherent responses. Experimental design questions are addressed in the next section. All three free-response question types are used in this book.

Since you are not given specific formulas for some concepts, you should begin learning how these formulas are derived starting at the beginning of the year. For example, you are not given the specific formulas for projectile motion problems since these are easily derived from the standard kinematics equations. If you begin reviewing a few weeks before the AP exam, you may want to make index cards of formulas to help you to memorize them.

You must read the entire question carefully before you begin. As you begin to solve the problem, make sure that you write down the general concept being used, for example, conservation of mechanical energy or conservation of energy. Then, you must write down the equations you are using. For example, if the problem requires you to use conservation of mechanical energy (potential and kinetic energies), write out those equations:

Initial total mechanical energy = Final total mechanical energy

$$mgh_i + \frac{1}{2}mv^2_i = mgh_f + \frac{1}{2}mv^2_f$$

When you are making substitutions, you must include the units! For example, if you are calculating net forces on a mass (such as a 2 kg mass that has an acceleration of 4 m/s^2), you must write as neatly as possible:

$$\Sigma F = F_{net} = ma = (2 \text{ kg})(4 \text{ m/s}^2) = 8 \text{ N}$$

Include all relevant information. Communicate with the grader by showing him/her that you understand what the question is asking. You may want to make a few sketches or write down your thoughts in an attempt to find the correct solution path. When a written response is requested, make sure that you write neatly and answer the question in full sentences.

Sometimes the question refers to a lab experiment typically performed in class or to simulated data that are given. In that case, you may be asked to make a graph. Make sure the graph is labeled correctly (with axes labeled and units clearly marked), points plotted as accurately as possible, and best-fit lines or curves used. Do not connect the dots. Always use the best-fit line for calculating slopes. Make sure you include your units when calculating slopes. Always show all of your work.

If you are drawing vectors, make sure the arrowheads are clearly visible. For angles, there is some room for variation.

Since angles are measured in degrees, be sure your calculator is in the correct mode. If scientific notation is used, make sure you know how to input the numbers into your calculator correctly. Remember, each calculator is different.

If you are asked to draw a free-body diagram (see pages 86–88), make sure you include only actual applied forces. Do not include component forces. Centripetal force is not an applied force and should not be included on a free-body diagram.

What do you do if you are not sure how to solve a problem? Follow these 11 tips.

1. Make sure you understand the general concepts involved and write them down.
2. Write down all appropriate equations.
3. Try to see how this problem may be similar to one you may have solved before.
4. Make sure you know which information is relevant and which information is irrelevant to what is being asked.
5. Rephrase the question in your mind. Maybe the question is worded in a way that is different from what you are used to.

TIP

Make sure you show all of your work on Part II. Include all formulas, substitutions with units, and general concepts used. Remember to label all diagrams. Communicate with the grader!

TIP

Make sure you have pencils, pens, a calculator, extra batteries, and a metric ruler with you for the exam!

6. Draw a sketch of the situation if one is not provided.

7. Write out what you think is the best way to solve the problem. This sometimes triggers or cues a solution.

8. Use numbers or estimations if the solution is strictly algebraic manipulation, such as deriving a formula in terms of given quantities or constants.

9. Relax. Sometimes if you move on to another problem, take a deep breath, close your eyes, and relax for a moment, the tension and anxiety may go away and allow you to continue.

10. Do not leave anything out. Unlike on the multiple-choice questions, you need to show all of your work to earn credit.

11. Understand what you are being asked to do. The Physics 1 exam wants you to respond in specific ways to certain key words.

   - "Justify" or "Explain": Support your answer with words, equations, calculations, diagrams, or graphs.
   - "Calculate": Provide numerical and algebraic work leading to the final answer.
   - "What is" or "Determine": Although showing work is always desirable, it is not necessary for these questions—you may simply state your answer.
   - "Derive": Starting with a fundamental equation (such as those given on the formula sheet), mathematically manipulate it to the desired form.
   - "Sketch": Without numerical scaling or specific data points, draw a graph that captures the key trend in the relationship (curvature, asymptotes, and so on).
   - "Plot": Specific data points should be placed onto a scaled grid. Do not connect the dots (although trends, especially linear ones, may be superimposed on the graph).

EXPERIMENTAL DESIGN QUESTIONS

According to the College Board, there are seven "Science Practices" that students should be familiar with when taking the exam. Although you need to know these practices throughout the exam, pay special attention to these when answering the experimental design questions.

1. **Appropriate use of representations and models:** Use diagrams, graphs, and equations when explaining the problem.

2. **Appropriate use of mathematics:** Define your variables, show algebraic manipulations clearly, and plug in only specific data at the very end of the problem. Do not use more significant digits in your answer than you have in the raw data.

3. **Scientific questioning:** All proposals must be testable and easy to understand. Phrase proposals in terms of specific relationships, such as "y will vary inversely with increasing x-values."

4. **Planning and implementing data collection strategies:** Systematic testing of one possible factor at a time while holding other factors constant is the key here.

5. **Data analysis and evaluations:** Graph the data when possible, and explicitly address whether or not the data are correlated (one variable has an effect on the other). Is there a linear trend? What is the line of best fit?

6. Work with scientific explanations and theories.

7. Connect and relate knowledge from various concepts, scales, and representations.

Fold your existing knowledge of physics in to your experimental design. Can you make analogies to other areas of physics to guide your thinking? ("Like force causes linear acceleration; the torques here will cause angular accelerations.")

GRAPHS, FITS, AND THE LINEARIZATION OF DATA

Graphs that are linear in nature are much easier to analyze, especially by hand, than graphs of any other nature. Trends, slopes, intercepts, and correlations of experimental results to theoretical predications are readily obtained. For this reason, if you are asked to graph your data, it will almost always be advantageous to linearize it first. Specifically, if the relationship is not linear to start, use a change of variable to make the relationship linear.

For example, if asked to determine the spring constant k of a system based on a collection of elastic potential energies for various extensions of the spring, the relevant equation is

$$\text{EPE} = \frac{1}{2}kx^2$$

This is a quadratic relationship, not a linear one. Before graphing, make the following change of variable:

$$z = x^2$$

So the relationship is now:

$$\text{EPE} = \frac{1}{2}kz$$

Now when graphed and a line of best fit is applied, the slope of the straight line will be ½k.

How can a line of best fit be generated by hand? If need be, you can use a straightedge and draw one straight line that has as many data points above the line as below it. Once this line is drawn, all subsequent calculations should be based on the slope and intercept of this best-line fit rather than on the original data. The idea here is that the fit of the data is an average of the raw data and is inherently better than any one particular point because the random variations in data have been smoothed out by the fit. When asked to analyze a graph of data, always use the fitted line or curve rather than the individual data points for this same reason.

Determine k by graphing the following data provided by another student.

EPE (J)	x (cm)
0.058	2.5
0.196	4.6
1.117	11.0
2.081	15.1

Solution

Since the relationship between these variables is quadratic, begin by squaring the given values for x. Also change to the standard MKS units of meters. Note that you must scale the axis yourself based on the given data. Your scale should make sure that at least two-thirds of the graph space provided is used.

x (cm)	x^2 (m^2)
2.5	0.000625
4.6	0.002116
11.0	0.0121
15.1	0.022801

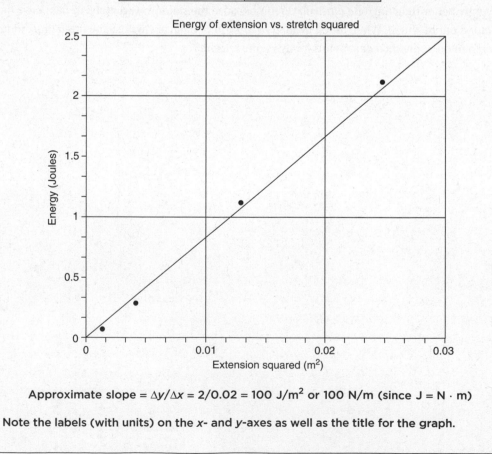

Approximate slope = $\Delta y / \Delta x$ = 2/0.02 = 100 J/m^2 or 100 N/m (since J = N · m)

Note the labels (with units) on the x- and y-axes as well as the title for the graph.

UNCERTAINTY AND PERCENT ERROR

In addition to respecting the number of significant digits in measured or recorded data, you can also determine the corresponding uncertainty in derived quantities. For example, if the radius of a circle is measured and recorded as 3.5 cm, there are only 2 significant digits in this number. Therefore, the area of the circle (πr^2) should be truncated from the calculator result to 2 significant digits. The area is 38 cm^2. The rest of the digits are not significant as they imply a precision in the radius that we do not have.

To take this analysis one step further, a percent error can be associated with a measurement. For example, one could write down the uncertainty in radius explicitly as $r = 4.5 +/- 0.05$ cm. This implies that the true value is most likely between 4.45 and 4.55 cm. This is a percent error of 1.1%.

$$\frac{0.05}{4.5} \times 100 = 1.1\%$$

$$100 * \text{Uncertainty} / \text{Value} = \text{Percent error}$$

Percent errors are an easier way of comparing the relative precision of different measurements. For example, a measurement of $128 +/- 2$ mm has a percent error of 1.6% and is thus less precise than our original measurement that has an error of 1.1%.

The College Board offers the following clarification about their expectations on uncertainty: "On the AP Physics 1 exam, students will not need to calculate uncertainty but will need to demonstrate understanding of the principles of uncertainty." Although a student should be able to use, reason, and support his or her answers by proper use of significant digits and percentage error, the student will not be expected to propagate errors, calculate standard deviations, or carry out formal linear regressions.

THE PARAGRAPH-LENGTH RESPONSE

According to the College Board, you should be "able to both analyze a situation and construct a coherent, sequenced, well-reasoned exposition that cites evidence and principles of physics and that make sense on the first reading." Points may be deducted for responses consisting primarily of equations, extraneous information, or arguments not presented in the correct logical order. The paragraph should not be overly long and may include equations, diagrams, graphs, and calculations; however, it should consist mostly of prose.

STUDY SKILLS AND SCHEDULING YOUR REVIEW

Preparing for any Advanced Placement exam takes practice and time. Effective studying involves managing your time so that you efficiently review the material. Do not cram a few days before the exam. Getting a good night's sleep before the exam and having a good breakfast the day of the exam is a better use of your time than "pulling an all-nighter." Working in a study group is a good idea. Using index cards to make your own flash cards of key concepts, units, and formulas can also be helpful.

When you study, try to work in a well lighted, quiet environment, when you are well rested. Studying late at night when you are exhausted is not an effective use of your time. Although some memorization may be necessary, physics is best learned (and studied) by actively solving problems. Remember, if you are using this book during the year, working through the chapter problems as you

> ## CONVENTIONS
>
> The following conventions are used on the AP Physics 1 exam and in this book:
>
> - The frame of reference of any problem is assumed to be inertial (unless otherwise stated).
> - Assume air resistance is negligible (unless otherwise stated).
> - In all situations, positive work is defined as work done on a system.
> - The direction of current is the direction in which positive charges would drift.
> - Assume all batteries and meters are ideal (unless otherwise stated).

cover each topic in class, memorizing the units, and familiarizing yourself with the formulas at that time will make your studying easier in the days before the exam.

If you are using this book in the weeks before the exam, make sure you are already familiar with most (if not all) of the units, equations, and topics to be covered. You can either use the chapter review for a quick overview and practice, or dive right in to the diagnostic test. You do not need to take the diagnostic test under timed conditions. See how you do, and then review the concepts for those questions that you got wrong. You can use the end-of-chapter questions to test your grasp of specific topics and then work on the remaining practice tests.

Setting up a workable study schedule is also vital to success. Each person's needs are different. The following schedule is just one example of an effective plan.

> Don't forget that you have *two* full-length practice tests with this book, in addition to the diagnostic test.

THE SEVEN BIG IDEAS

The College Board has identified seven "Big Idea" strands that serve as the central organizing structure for both the AP Physics 1 and AP Physics 2 exams. Although the tests have different content, they both attempt to reinforce these central ideas in ways that are appropriate to the subject matter.

- Big Idea 1: "Objects and systems have properties such as mass and charge. Systems may have internal structure."
- Big Idea 2: "Fields existing in space can be used to explain interactions." (AP Physics 1 has only gravitational fields.)
- Big Idea 3: "The interactions of an object with other objects can be described by forces."
- Big Idea 4: "Interactions between systems can result in changes in those systems."
- Big Idea 5: "Changes that occur as a result of interactions are constrained by conservation laws."
- Big Idea 6: "Waves can transfer energy and momentum from one location to another without the permanent transfer of mass and serve as a mathematical model for the description of other phenomena."
- Big Idea 7: "Probability can be used to describe quantum systems." (AP Physics 2 only.)

Use these Big Idea strands as a guide when studying for the tests. They indicate what conceptual questions may appear on the exam.

OBJECTS VERSUS SYSTEMS

An **object** is thought of as an isolated mass that is being acted upon by outside forces. Any internal structure of the object is ignored. The object is defined by its properties (e.g., mass, charge). Strictly speaking, for a single object being analyzed, it does not make sense to make use of potential energy, Newton's third law, thermal energy, or the conserved quantities. Instead when discussing a single object, use the following relationships from mechanics:

$$\vec{F}_{net} = m\vec{a}$$

$$\vec{\tau} = I\vec{\alpha}$$

$$\vec{F}\Delta t = m\Delta\vec{v}$$

$$F\Delta d = \Delta KE$$

A **system**, in contrast, is made up of objects that may interact. It is within isolated systems that the conserved quantities, canceling Newton's third law forces (internal forces), the heat capacity of the constituent parts, and the potential energy of relationship between interacting objects are useful concepts. Within systems, conserved quantities become the most useful lens through which to view the situation:

$$\sum\vec{p}_i = \sum\vec{p}_f$$

$$\sum\vec{L}_i = \sum\vec{L}_f$$

$$ME_i = ME_f$$

Table 3. Test Prep Schedule

September 1–April 15	As the year progresses, make sure you memorize units and are comfortable with formulas. If you are using this book during the year, do end-of-chapter problems as they are covered in class. Make sure you register for the exam, following school procedures, and refer to the College Board's website for details: *www.collegeboard.com*
Four weeks before the exam	Most topics should be covered by now in class. If you are using this book for the first time, begin reviewing concepts and doing the end-of-chapter problems. Begin reviewing units and formulas. Devote at least 30 minutes each day to studying.
Three weeks before the exam	Start working on the diagnostic test. Go back and review topics that you are unsure of or feel that you answered incorrectly.
Two weeks before the exam	Begin working on practice tests. Continue to review old concepts.
One week before the exam	Do the remaining practice tests timed. Make sure you are comfortable with the exam format and know what to expect. Review any remaining topics and units.
The day before the exam	Pack up your registration materials, pens, pencils, calculator, extra batteries, and metric ruler. Put them by the door, ready to go. Get a good night's sleep.
The day of the exam	Have a good breakfast. Make sure you take all the items you prepared the night before. Relax!

SUMMARY

- Make sure you set up a manageable study schedule well in advance of the exam.
- Make sure you memorize all units and are familiar with the exam format.
- Multiple-choice questions do not have a penalty for wrong answers, so do not skip any. If you are unsure of the answer, try to eliminate as many choices as you can, and then guess!
- Do not leave any question out on the free-response part! Show all of your work. Write down all fundamental concepts, write all equations used, and include units for all substitutions and in your final answer.
- Read each question carefully. Write your answers clearly. On the multiple-choice questions, make sure you have a #2 pencil and bubble in all information carefully. Write out short-answer questions in full sentences. Clearly label graphs with units and use best-fit lines or curves.
- Know the "Seven Big Ideas" of physics as defined by the College Board; they are helpful clues as to what key concepts are emphasized on the exam.
- An object can be represented by a single mass. When exposed to external forces, it is best modeled as experiencing changes in speed and/or direction.
- A system is a group of objects which, if isolated, is best modeled with conservation laws.
- Try to relax and do all of the practice tests. Work on the chapter questions to review concepts as needed.
- Get a good night's sleep before the exam.
- On the day of the exam, bring all registration materials with you, as well as pens, pencils, calculators, extra batteries, and a metric ruler.

Relax and Good Luck!

Diagnostic Test

This section contains a short diagnostic test. The purpose of this diagnostic exam is for you to identify those conceptual areas most in need of review. The relevant sections of the book (indicated in the answer keys) should be reviewed thoroughly before attempting one of the full-length practice tests at the end of this book.

Diagnostic Test

AP Physics 1

SECTION I: MULTIPLE-CHOICE

DIRECTIONS: Select the best answer (or two best answers when indicated) on each of the following 26 multiple-choice questions as well as the 2 free-response questions. The detailed answer explanations will direct you to a specific chapter for further review on the specific subject matter in that question. You may use a calculator and make use of the formula sheet provided in the appendix.

1. A mass is suspended from the ceiling by one rope at an angle as indicated. The mass is held in place by a second, horizontal rope. Without knowing the exact angle from the ceiling, except that it is greater than 45 degrees, which statement best represents the possible value for tension in the horizontal line?

   (A) $T > mg$
   (B) $T = mg$
   (C) $T < mg$
   (D) cannot be determined without knowing the angle

2. Pushing on a mass of 15 kg with a sideways force of 120 N is not enough to get the mass to start sliding across the floor. Which of the following best describes the coefficient of static friction between the floor and the mass?

   (A) $\mu = 0.8$
   (B) $\mu > 0.8$
   (C) $\mu < 0.8$
   (D) $\mu = 1.25$

3. While standing on an elevator and experiencing a downward acceleration of 4.9 m/s^2, what is the force between the floor and your feet? The variable m is your mass.

(A) mg

(B) $0.5mg$

(C) $1.5mg$

(D) $4.9mg$

4. After throwing a ball at an upward angle, the reason the ball continues to go horizontally while falling due to gravity is that

(A) the ball's inertia keeps it going

(B) the force of the throw keeps it going

(C) the energy from gravitational potential keeps replenishing the lost energy

(D) air pressure keeps the ball from falling too quickly

5. Consider the situation in which your hand is pushing a book across a rough desktop with lots of friction. Of the many forces involved, consider only these two individual forces: the force on your hand from the book and the force the book is experiencing from your hand. While the book is accelerating from rest to its final velocity, which statement best compares the force experienced by your hand compared with that experienced by the book?

(A) Force on book > force on hand

(B) Force on hand > force on book

(C) Force on hand = force on book

(D) The relationship between these two forces depends on the frictional force.

6. A puck comes to a stop on a level floor in 20 meters after an initial speed of 10 m/s. What is the coefficient of kinetic friction between the puck and the floor?

(A) The mass is required to calculate this answer.

(B) 0.25

(C) 0.5

(D) 0.025

7. A rock is dropped from an extremely high cliff and experiences free-fall conditions. Which of the following statements is NOT true about the rock's velocity, acceleration, or displacement between the 3rd and 4th seconds of falling?

(A) The rock will fall an additional 4.9 meters.

(B) The rock will speed up by 9.8 m/s.

(C) The rock's acceleration will remain 9.8 m/s^2.

(D) The rock's displacement, velocity, and acceleration vectors are all directed downward.

8. When using the kinematics equation $x = x_0 + v_{x0}t + \frac{1}{2}a_x t^2$, which of the following is always assumed to be true?

(A) $a_x = $ zero

(B) $x > x_0$

(C) V_{x0} is positive.

(D) a_x is constant.

9. Consider a projectile launched at 30 degrees above a flat surface. The projectile experiences no air resistance and lands at the same height from which it was launched. Compare the initial and final conditions of the following vector components:

$$V_x \quad V_y \quad a_x \quad a_y$$

(A) They are all the same at both instances.
(B) They are all different from their initial values.
(C) V_x and V_y have changed, but a_x *and* a_y are constant.
(D) They are all the same except for V_y.

10. A physics teacher is swirling a bucket around in a vertical circle. If the bucket slips out of his hand when it is directly overhead, the bucket will

(A) fly upward initially
(B) drop straight down
(C) fly out horizontally initially
(D) fly out up and at an angle initially

11. If the Moon was twice as far from the center of Earth as it is currently, how would its orbital period change?

(A) The Moon's orbital period would double.
(B) The Moon's orbital period would increase by a factor of 4.
(C) The Moon's orbital period would increase by a factor of $2\sqrt{2}$.
(D) The Moon's orbital period would decrease by a factor of 2.

12. If you went to another planet, which of the following would be true?

(A) Your mass would be the same, but your weight would change.
(B) Your mass would change, but your weight would be the same.
(C) Your mass would be the same, and your weight would be the same.
(D) Your mass would change, and your weight would change.

13. You lift a heavy suitcase twice. Each time you lift it to the same height, but the second time you do it in half the time. Compare the work done to and power delivered to the suitcase.

(A) Same work, same power
(B) Twice the work, twice the power
(C) Twice the work, same power
(D) Same work, twice the power

14. A constant horizontal force of 12 N is applied for 5 meters to move an initially stationary mass of 5 kg. The friction between the floor and the mass does a total of −40 joules of work to the mass. The final speed (in m/s) of the object is

(A) $2(6)^{\frac{1}{2}}$
(B) $2(2)^{\frac{1}{2}}$
(C) $2(10)^{\frac{1}{2}}$
(D) 4

15. A skier begins a downhill run at a height of 9 meters and a speed of 2 m/s. If you ignore friction, what will be her speed when she is 3 meters from the bottom of the slope?

    (A) 6.0 m/s
    (B) 12.8 m/s
    (C) 11.1 m/s
    (D) 7.7 m/s

16. A rubber ball ($m = 0.250$ kg) strikes a wall horizontally at 3.50 m/s and rebounds elastically. What is the magnitude of impulse delivered to the wall in N · s?

    (A) 0.875
    (B) 1.75
    (C) 2.50
    (D) 8.75

17. A running back ($m = 85$ kg) running at 1.5 m/s is tackled from the side by another player ($m = 75$ kg) running perpendicularly to the running back's original heading at 1.75 m/s. What is the resulting speed of the two entangled players just after the tackle?

    (A) 2.2 m/s
    (B) 0.25 m/s
    (C) 1.6 m/s
    (D) 1.1 m/s

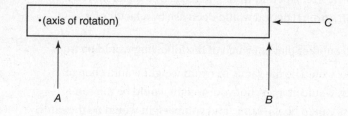

18. Which of the following ranks the torques applied by the three equal forces A, B, and C to the long thin rod about its fixed axis as indicated?

    (A) $\tau_C > \tau_B > \tau_A$
    (B) $\tau_A = \tau_B > \tau_C$
    (C) $\tau_B = \tau_C > \tau_A$
    (D) $\tau_B > \tau_A > \tau_C$

19. A large spherical cloud of dust in deep space that is spinning slowly collapses under its own gravitational force into a spherical cloud 10 times smaller in diameter. How does its rotational speed change?

    (A) The rotational speed does not change.
    (B) The rotational speed increases by a factor of 10.
    (C) The rotational speed decreases by a factor of 10.
    (D) The rotational speed increases by a factor of 100.

20. A single resistor attached to a constant voltage source has an additional, identical resistor inserted between one end of the resistor and the voltage supply. Which of these is the correct description of what will happen to the voltage drop across (V) and the current running through the original resistor (I) as a result?

   (A) I and V will remain unchanged.
   (B) I and V will both be halved.
   (C) I will be halved; V will remain unchanged.
   (D) V will be halved; I will remain unchanged.

21. A single resistor attached to a constant voltage source has an additional, identical resistor inserted beside it, forming its own connection across the voltage supply. Which of these is the correct description of what will happen to the voltage drop across (V) and the current running through the original resistor (I) as a result?

   (A) I and V will remain unchanged.
   (B) I and V will both be halved.
   (C) I will be halved; V will remain unchanged.
   (D) V will be halved; I will remain unchanged.

22. A simple circuit consisting of a battery, a switch, and a lightbulb in series is constructed. Which is the best explanation of what happens when the switch is closed for the first time?

   (A) The lightbulb goes out immediately.
   (B) Electrons are allowed to flow through the switch from the battery. When they reach the bulb, it lights.
   (C) Electrons are allowed to flow through the switch. When those electrons already within the bulb move, the lightbulb lights.
   (D) Electrons on the negative side of the switch are allowed to first flow. When they reach the bulb, it lights.

23. As two sine waves run into each other on a single rope under tension, as shown above, what is the resulting sequence of events?

   (A) Initial constructive interference followed by unchanged waves emerging on opposite sides
   (B) Initial constructive interference followed by waves bouncing backward
   (C) Initial partial destructive interference followed by constructive interference followed by unchanged waves emerging on opposite sides
   (D) Initial partial destructive interference followed by constructive interference followed by waves bouncing backward

24. As you stand on the side of the road, an ambulance approaches your position at high speed. Which option best describes the sound of the siren as observed by you as compared with the driver's perception?

	Frequency	Wavelength	Wave Speed
(A)	Higher	Lower	Same
(B)	Higher	Same	Same
(C)	Higher	Same	Higher
(D)	Lower	Higher	Higher

25. Two cars approach an intersection at right angles, as pictured above. What is their relative speed to each other?

(A) 10 mph

(B) 45 mph

(C) 57 mph

(D) 80 mph

26. Which of the following statements are true regarding collisions? Select two answers.

(A) An elastic collision conserves both net momentum and net kinetic energy.

(B) An inelastic collision conserves mechanical energy but not net momentum.

(C) An elastic collision between two objects has equal but opposite impulses delivered.

(D) An inelastic collision between two objects is one in which both objects come to rest during the collision.

SECTION II: FREE-RESPONSE

1. A static, nonuniform wedge-shaped mass, m (with center of mass x as indicated), with a base length of L (shown below) is supported at two points. The first support point is at the bottom left corner of the wedge. The second support point is located at a point ¾ of the length of the wedge away from that corner.

(a) Using as givens m, L, and g (the gravitational field strength), solve for the two unknown contact forces at the points of support (F_1 and F_2).

(b) Support #2 is removed. The wedge begins to fall by rotating around support #1 with an angular acceleration of 2 rad/s². If the mass of the wedge is 3.8 kg and $L = 75$ cm, determine the moment of inertia of the nonuniform wedge about the top of support #1.

(c) As the wedge continues to rotate clockwise but remains in contact with support #1:

(i) Does the angular acceleration increase, decrease, or remain the same? Justify your response.

(ii) Does the moment of inertia of the wedge increase, decrease, or remain the same? Justify your response.

2. An object of unknown mass is fired at an unknown angle with an initial speed of 25 m/s. Ignore frictional effects. If only 45 percent of its initial kinetic energy is still present as kinetic energy at the projectile's highest point, determine the time in flight for the projectile.

ANSWER KEY	Topic and Chapter to Reference
1. **(C)**	Statics problem, Chapter 3
2. **(B)**	Frictional dynamics, Chapter 3
3. **(B)**	Elevator problem, Chapter 3
4. **(A)**	Newton's first law, Chapter 3
5. **(C)**	Newton's third law, Chapter 3
6. **(B)**	Friction, Chapters 2 and 3
7. **(A)**	Friction, Chapters 2 and 3
8. **(D)**	Kinematics, Chapter 2
9. **(D)**	Projectile motion, Chapter 2
10. **(C)**	Circular motion, Chapter 3
11. **(C)**	Gravity, Chapter 5
12. **(A)**	Mass vs. weight, Chapter 3
13. **(D)**	Work and power, Chapter 4
14. **(B)**	Net work = ΔKE, Chapter 4
15. **(C)**	Net work = ΔKE, Chapter 4
16. **(B)**	Impulse and momentum, Chapter 6
17. **(D)**	Conservation of momentum, Chapter 6
18. **(D)**	Torque, Chapter 7
19. **(D)**	Conservation of angular momentum, Chapter 7
20. **(B)**	Series circuits, Chapter 9
21. **(A)**	Parallel circuits, Chapter 9
22. **(C)**	Open/closed circuit, Chapter 9
23. **(C)**	Open/closed circuit, Chapter 9
24. **(A)**	Sound waves, Chapter 10
25. **(C)**	Relative velocity, Chapter 2
26. **(A)** and **(C)**	Conservation of linear momentum, Chapter 6

ANSWERS EXPLAINED

Section I: Multiple-Choice

1. **(C)** $T < mg$. Since the mass is stationary, we conclude that the net force is zero. The upward tension of the angled rope must be mg in order to cancel out gravity. The free-body diagram on the hanging mass is:

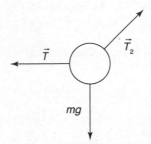

Since the angle θ is greater than 45 degrees, the vertical component of tension T_2 must be bigger than the horizontal component. (To see this quickly, draw an extreme case, such as 85 degrees.) Therefore, the horizontal component of tension must be less than mg. The horizontal rope T must cancel this horizontal component of T_2.

2. **(B)** $\mu > 0.8$. The normal force between the mass and the floor is 150 N. The needed sideways push to overcome static friction is **greater than** 120 N. Dividing these two forces would give us:

$$\mu_s = 120/150 = 0.8$$

Since a greater force is needed, the coefficient of static friction must be greater than this value.

3. **(B)** $0.5mg$. The free-body diagram consists of only two forces.

Newton's second law gives us $+N - mg = ma$. Let $a = -0.5g$, which is half the value of g and downward:

$$N - mg = -0.5mg$$

$$N = 0.5mg$$

4. **(A)** The ball's inertia keeps it going. Newton's first law says that a body's inertia maintains its current motion once in motion. After the force of the throw, the only force on the ball is the vertical force of gravity. This causes the ball to accelerate in the vertical direction while the ball's sideways velocity is maintained due to its inertia.

5. **(C)** Force on hand = force on book. Newton's third law of action-reaction states that the force pair of interaction between two objects is always equal and opposite. The acceleration of either object has no role in the third law. The acceleration comes about from the net force on a single object.

6. **(B)** 0.25. To find force, we must first find acceleration:

$$v_f^2 = v_i^2 + 2ad$$

$$(0)^2 = (10 \text{ m/s})^2 + (2a)(20 \text{ m})$$

$$a = -2.5 \text{ m/s}^2$$

The net force horizontally is the friction. There are no other horizontal forces:

Newton's second law gives us:

$$f = ma = -m(2.5)$$

Our model for friction gives us:

$$f = -\mu N$$

This value is negative since it is to the left. Since the vertical forces cancel out, $N = mg$. Substituting for friction from above and mg for N:

$$-m(2.5) = -\mu mg$$

Note that mass cancels:

$$\mu = (2.5/g) = 0.25$$

7. **(A)** The rock will fall an additional 4.9 meters is NOT true. An object in free fall will accelerate by a constant 9.8 m/s². Since the rock is already moving downward, this leads to an increase in speed of 9.8 m/s in that 1-second interval. However, the rock will fall much farther than 4.9 m as it is already moving close to 30 m/s downward by the start of the 3rd second. It will fall an additional 35 meters in that 4th second. The rock does indeed fall 4.9 meters in the 1st second of the drop but falls an increasingly greater distance each second thereafter.

8. **(D)** a_x is constant. All the equations of motion are predicated on the assumption of constant acceleration. Any permutation of signs or values will work in any of the equations of motion. If the acceleration is not constant, one must use graphical methods or approximate the acceleration as constant over a narrow interval of time.

9. **(D)** They are all the same except for V_y. Projectile motion involves a constant vertical acceleration of $-g$ and no horizontal acceleration at all. Therefore, a_x and a_y are both constant throughout the flight. Since a_x is zero, there is no change in V_x. So V_x is also constant. V_y, however, is always changing. For a symmetric problem such as this one, the projectile lands with the same vertical speed as it had initially but in the opposite direction so that $V_{yf} = -V_{yi}$.

10. **(C)** The bucket will fly out horizontally initially. Objects executing circular motion have velocity vectors tangent to their motion while experiencing an inward acceleration. If the inward force is released, Newton's first law dictates that the object follow its velocity vector. Note that gravity will cause the object to fall after it is released, but the initial motion is horizontal.

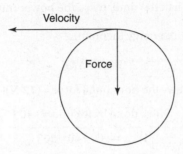

11. **(C)** The Moon's orbital period would increase by a factor of $2\sqrt{2}$. Kepler's third law of orbital bodies states that their orbital periods squared are proportional to their orbital radius cubed:

$$T_{old}^2 / R_{old}^3 = T_{new}^2 / R_{new}^3$$

Substituting $R_{new} = 2R_{old}$:

$$T_{old}^2 / R_{old}^3 = T_{new}^2 / (2R_{old})^3$$

Cancel R and solve for T_{new}:

$$T_{new} = 8^{1/2} T_{old} = 2\sqrt{2} T_{old}$$

Note that you can also obtain this result by examining the equation for orbital speed from Newton ($v = (Gm/r)^{1/2}$). This shows that the orbital speed of the Moon must decrease by the square root of 2. Coupling this with the knowledge that the new orbit is now twice as large, it will take $2(2^{1/2})$ or $2\sqrt{2}$ times as long to orbit once around Earth as it used to.

12. **(A)** Your mass would be the same, but your weight would change. Mass is a property of an object and measures the object's inertia regardless of location. Weight is a force due to gravity and depends both on the object's mass and on the planetary gravity the object is currently experiencing.

13. **(D)** Same work, twice the power.

$$\text{Work} = \text{force} \times \text{distance}$$

For lifting operations, the force is assumed to be equal to the weight of the object for most of the lift. (Briefly, at the beginning of the lift, the lifting force must be greater than the weight in order to get the object moving. This brief force must be different in the second case in order to get the suitcase moving faster. However, this initial force is generally ignored in these calculations as it does not happen over an appreciable distance, just as the decrease in lifting force at the end of the lift is likewise ignored.) The same force over the same distance yields the same work. If the details of the forces contributing to the work are troublesome, you may prefer to think about the energy gained by the suitcase ($W = \Delta E$). Since the suitcase has gained the same amount of energy in both cases by being lifted, the work done must likewise be the same:

$$\text{Power} = \text{work/time}$$

If the same work is done in half the time, twice the power must be supplied.

14. **(B)** $2(2)^{\frac{1}{2}}$ m/s. From the work-energy theorem:

$$W_{\text{net}} = \Delta KE$$

$$\text{Work done by the horizontal force} = (12 \text{ N})(5 \text{ m}) = 60 \text{ J}$$

$$\text{Work done by friction} = -40 \text{ J}$$

$$W_{\text{net}} = +60 - 40 = 20 \text{ J}$$

So the kinetic energy of the mass has gone up by 20 J. Initially, it had no KE since it was stationary. Therefore:

$$\frac{1}{2}mv_f^2 = 20$$

$$v_f^2 = 40/5 = 8$$

$$v_f = 2(2^{\frac{1}{2}})$$

15. **(C)** 11.1 m/s. Mechanical energy is conserved since there are no nonconservative forces involved:

$$ME_i = ME_f$$

$$KE_i + PE_i = KE_f + PE_f$$

$$\frac{1}{2}m(2 \text{ m/s})^2 + mg(9 \text{ m}) = \frac{1}{2}mv_f^2 + mg(3 \text{ m})$$

Canceling mass in every term and solving for v_f yields:

$$v_f^2 = 2^2 + 2 \times (9g - 3g) = 4 + 12g = 124$$

$$v_f = 11.1 \text{ m/s}$$

16. **(B)** 1.75

$$\text{Impulse} = \Delta\text{momentum}$$

$$= m\vec{v}_f - m\vec{v}_i$$

Since the collision is elastic, no kinetic energy is lost. Assuming the wall remains stationary, the rubber ball must rebound with the same speed but in the opposite direction:

$$\vec{v}_f = -\vec{v}_i$$

Substituting for v_f above:

$$\text{Impulse} = m(-\vec{v}_i) - m\vec{v}_i = -2m\vec{v}_i = -2(0.25\text{ kg})(3.5\text{ m/s}) = -1.75\text{ kg}\cdot\text{m/s}$$

Here the negative sign indicates that the impulse delivered to the ball is in the opposite direction from the ball's initial velocity. Note the ball and the wall receive equal and opposite impulses (from Newton's third law) and the problem asks for magnitude (absolute value).

Since $F\Delta t$ is also an impulse, units of N · s must be the same as kg · m/s and is quickly shown:

$$\text{N}\cdot\text{s} = (\text{kg}\cdot\text{m/s}^2)\text{s} = \text{kg}\cdot\text{m/s}$$

17. **(D)** 1.1 m/s. Total momentum must be conserved through the collision. So we need to find only the magnitude of the initial momentum of the system. Each player is originally running perpendicularly to the other. Thus, their individual momentums are different components. Imagine one as an x-component and the other as a y-component. Therefore, the resulting magnitude is found using the Pythagorean theorem. Dividing by total mass yields the speed of the combined players after impact:

$$p_{1x} = 85(1.5) = 127.5\text{ kg}\cdot\text{m/s}$$

$$p_{2y} = 75(1.75) = 131.25\text{ kg}\cdot\text{m/s}$$

$$\text{Magnitude} = 183\text{ kg}\cdot\text{m/s}$$

$$\text{Net mass} = 160\text{ kg}\cdot\text{m/s}$$

$$\text{Speed} = \text{magnitude/mass} = 1.14\text{ m/s}$$

18. **(D)** $\tau_B > \tau_A > \tau_C$

$$\text{Torque} = (\text{force})(\text{distance from axis of rotation})\sin\theta$$

Force C is applied along the lever arm. So force C has an angle of 180°, resulting in no torque.

Force A has a much shorter lever arm than force B. Therefore, the torque supplied by A is much less than the torque supplied by B.

19. **(D)** The rotational speed increases by a factor of 100. The moment of inertia (I) for a sphere is proportional to R^2. Because the shape of the cloud is not changing, we can determine that the moment of inertia has decreased by a factor of 10^2. Since angular momentum is conserved:

$$I_1\omega_1 = I_2\omega_2$$

The new angular velocity (ω_2) must be 100 times bigger:

$$I_1\omega_1 = I_2\omega_2 = (I_1/100)(100\omega_1)$$

20. **(B)** I and V will both be halved. The effective resistance of the entire circuit is doubled, meaning the voltage supply will send out only half the current:

$$V = IR$$
$$V_{\text{same}}/R_{\text{doubled}} = I$$

The voltage used for the entire circuit must be shared between the two resistors. Since the two resistors are identical, the split will be 50/50.

21. **(A)** I and V will remain unchanged. The new resistor forms its own parallel path and therefore does not affect the voltage for the original path. Since the original resistor has the same voltage and resistance, it will draw the same current. Note, however, that the voltage supply must now furnish twice the current with the second resistor in place.

22. **(C)** Electrons are allowed to flow through the switch. When those electrons already within the bulb move, the lightbulb lights. Closing a switch completes the path, allowing electrons to move around the circuit. All components within the circuit have electrons that respond to the voltage difference and begin to flow. This resulting movement of electrons already within the bulb causes the lightbulb to glow. Recall that the drift velocities of electrons in simple circuits that make up the current are only about 0.03 cm/s.

23. **(C)** Initial partial destructive interference followed by constructive interference followed by unchanged waves emerging on opposite sides. After the waves superpose, they emerge as they were before as if they had never met. As they begin to overlap, the leading trough of the wave on the left will superpose with the leading peak of the wave on the right to interfere destructively. However, since the other halves of the waves are unaffected, this is only partial destruction. When the waves are completely superposed, trough will be on trough and peak on peak, leading to a constructive interference.

24. **(A)** Higher frequency, lower wavelength, same wave speed. Wave speed is determined by the medium, which is the air. So wave speed will not be observed to be different by either observer. Wavelength will appear to be shorter to the stationary observer as the crests are compressed by the ambulance moving toward you between peaks:

$$\text{Wavelength} \times \text{frequency} = \text{wave speed}$$

Since wavelength is shorter, the frequency must be higher in order to give the same speed. This higher frequency is the directly observable higher pitch that is known as the Doppler shift.

25. **(C)** 57 mph. Relative velocity is the difference in the velocity vectors. When the vectors are subtracted, the resultant is the hypotenuse of a right triangle:

$$\text{Relative velocity} = (35^2 + 45^2)^{\frac{1}{2}}$$

26. **(A)** and **(C)** An elastic collision conserves both net momentum and net kinetic energy. An elastic collision between two objects has equal but opposite impulses delivered. All collisions conserve net momentum. Elastic collisions also conserve kinetic energy. All collisions involve equal but opposite exchanges of momentums. Inelastic collisions simply indicate that kinetic energy is not conserved.

Section II: Free-Response

1. This is a statics problem (see Chapters 3 and 7 for more information).

(a) Both net force and net torque must add up to zero. Make a free-body diagram:

$$F_{net} = F_1 + F_2 - mg = ma = 0$$

$$F_1 + F_2 = mg$$

We now have two unknowns but only one equation! For torque, we are free to choose any axis of rotation we want since the wedge is not actually moving. We should place the axis of rotation at either contact point in order to eliminate one variable. If we place our axis of rotation at the F_1 contact point and use minus for clockwise rotations and plus for counterclockwise rotations:

$$\text{Net torque} = F_1(0) - mg(L/3) + F_2(3L/4) = I\alpha = 0$$

$$F_2(3L/4) = mg(L/3)$$

$$F_2 = 4mg/9$$

Combine this equation with the one found above.

$$F_1 = mg - 4mg/9 = 5mg/9$$

(b) $$\text{Torque} = I\alpha$$

Since F_2 is gone, the torque is supplied only by the mass of the object:

$$\text{Torque} = FR\sin\theta$$

$$\text{Torque} = mg(L/3)(1)$$

Substituting into the above equation:

$$mgL/3 = I\alpha = I(2)$$

$$I = mgL/6 = (3.8)(9.8)(0.75)/6 = 4.7 \text{ kg} \cdot \text{m}^2$$

Note that $I = mr^2$ cannot be used, as the object has mass spread out over many different r values.

(c) (i) As the object rotates, neither the lever arm nor the force (mg) changes. However, the angle between the lever arm and force goes from 90° initially toward 0° at the lowest point of the downward swing. Since torque is proportional to sin θ, the torque (and thus the angular acceleration) will decrease.

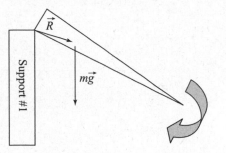

(ii) Since neither the distribution of mass nor the axis of rotation changes during the swing, the object's moment of inertia remains constant.

2. This is a projectile motion (Chapter 2) and energy conservation (Chapter 4) problem. Energy conservation tells us that the other 55 percent of the initial energy is in gravitational potential energy.

$$\text{GPE at top} = 55\% \text{ of initial } KE = 0.55\left(\frac{1}{2}m(25^2)\right)$$

$$mgh = 0.55\left(\frac{1}{2}m(25^2)\right)$$

Solve for height and note that mass cancels:

$$h = 17.5m$$

Determine the time to fall from that height. Remember there is no vertical velocity at the highest point!

$$D_y = \frac{1}{2}at^2$$

$$17.5 = 4.9t^2$$

$$t = 1.89 \text{ s}$$

Double this value to obtain time in flight: 3.8 seconds.

Review and Practice

Val2000

Vectors

KEY CONCEPTS
- → **COORDINATE SYSTEMS AND FRAMES OF REFERENCE**
- → **VECTORS**
- → **ADDITION OF VECTORS**
- → **SUBTRACTION OF VECTORS**
- → **ADDITION METHODS USING THE COMPONENTS OF VECTORS**

COORDINATE SYSTEMS AND FRAMES OF REFERENCE

We begin our review of physics with the idea that all observations and measurements are made relative to a suitably chosen frame of reference. In other words, when observations are made that will be the basis of future predictions, we must be careful to note from what point of view those observations are being made. For example, if you are standing on the street and see a car driving by, you observe the car and all of its occupants moving relative to you. However, to the driver and other occupants of the car, the situation is different: they may not appear to be in motion relative to themselves; rather, you appear to be moving backward relative to them.

If you were to get into your car, drive out to meet the other car, and travel at the same speed in the same direction right next to the first car, there would be no relative motion between the two cars. These different points of view are known as **frames of reference**, and they are very important aspects of physics.

A coordinate system within a frame of reference is defined to be a set of reference lines that intersect at an arbitrarily chosen fixed point called the **origin**. In the Cartesian coordinate system, the reference lines are three mutually perpendicular lines designated x, y, and z (see Figure 1.1). The coordinate system must provide a set of rules for locating objects within that frame of reference. In the three-dimensional Cartesian system, if we define a plane containing the x- and y-coordinates (let's say a horizontal plane), then the z-axis specifies direction up or down.

It is often useful to compare observations made in two different frames of reference. In the example above, the motion of the occupants of the car was reduced to zero if we transformed our coordinate system to the car moving at constant velocity. In this case, we say that the car is an **inertial frame of reference**. In such a frame, it is impossible to observe whether or not the reference frame is in motion if the observers are moving with it.

> **REMEMBER**
>
> A **frame of reference** represents an observer's viewpoint and requires a coordinate system to set the origin. A frame of reference moving with a constant velocity is called an **inertial frame of reference**.

Figure 1.1

VECTORS

Another way of locating the position of an object in the Cartesian system is with a directed line segment, or **vector**. If you draw an arrow, starting from the origin to a point in space (see Figure 1.2), you have defined a position vector $\vec{R}$, whose magnitude is given by $|\vec{R}|$ and is equal to the linear distance between the origin and the point (x, y). The direction of the vector is given by the angle, identified by the Greek letter θ, that the arrow makes with the positive x-axis. Any quantity that has both magnitude and direction is called a **vector quantity**. Any quantity that has only magnitude is a **scalar quantity**. Examples of vector quantities are force, velocity, weight, and displacement. Examples of scalar quantities are mass, distance, speed, and energy.

TIP

Vectors have both magnitude (size) and direction. Scalars have only magnitude.

Figure 1.2

If we designate the magnitude of the vector $\vec{R}$ as r, and the direction angle is given by θ, then we have an alternative coordinate system called the **polar coordinate system**. In two dimensions, to locate the point in the Cartesian system (x, y) involves going x units horizontally and y units vertically. Additionally, we can show that (see Figure 1.2)

$$x = r\cos\theta \quad \text{and} \quad y = r\sin\theta$$

Having established this polar form for vectors, we can easily show that r and θ can be expressed in terms of x and y as follows:

$$r^2 = x^2 + y^2$$

$$\tan\theta = \frac{y}{x}$$

The magnitude of $\vec{R}$ is then given by

$$|\vec{R}| = r = \sqrt{x^2 + y^2}$$

and the direction angle θ is given by

$$\theta = \tan^{-1}\frac{y}{x}$$

Here is an example of a vector described in two equivalent ways:

1. **MAGNITUDE AND DIRECTION:** useful for drawing, visualizing, and graphical analysis
2. **COMPONENT FORM:** useful for calculations and mathematical operations

You must be able to use the math in this section to go back and forth between these two representations fluently.

Magnitude and Direction	Component Form
Magnitude of $V = |\vec{V}| = 12$ units Direction of $V = \theta = 120$ degrees (standard angle) or 60 degrees above the –x-axis	Horizontal component of $V = V_x =$ $12 \cos(120°) = -12 \cos(60°) = -6$ units = 6 units to the left Vertical component of $V = V_y = 12 \sin(120°)$ $= +12 \sin(60°) = 10.4$ units $= 10.4$ units upward

ADDITION OF VECTORS

Geometric Considerations

The ability to combine vectors is a very important tool of physics. From a geometric standpoint, the "addition" of two vectors is not the same as the addition of two numbers. When we state that, given two vectors $\vec{A}$ and $\vec{B}$, we wish to form the third vector $\vec{C}$ such that $\vec{A} + \vec{B} = \vec{C}$, we must be careful to preserve the directions of the vectors relative to our chosen frame of reference.

> **REMEMBER**
>
> Vectors add constructively from head to tail. The vector sum is called the **resultant**.

One way to do this "addition" is by the construction of what is called a **vector diagram**. Vectors are identified geometrically by a directed arrow that has a "head" and a "tail." We will consider a series of examples.

EXAMPLE 1

First, suppose a girl is walking from her house a distance of five blocks east and then an additional two blocks east. How can we represent these displacements vectorially? First, we must choose a suitable scale to represent the magnitude lengths of the vectors. In this case, let us just call the scale "one vector unit" or just "one unit," to be equal to one block of distance. Let us also agree that "east" is to the right (and hence "north" is directed up). A vector diagram for this set of displacements is given in Figure 1.3. Note that the resultant has both magnitude and direction as it, too, is a vector.

5 blocks E   2 blocks E

Resultant = 7 blocks E

Figure 1.3

EXAMPLE 2

Now, for our second example, suppose the girl walks five blocks east and then two blocks northeast (that is, 45 degrees north of east). As before, we draw our vector diagram, using the same scale as before, so that the two vectors are connected "head to tail" or "tip to tail." The vector diagram will look like Figure 1.4. The resultant is drawn from the tail of the first vector to the head of the second, forming a triangle (some texts use the "parallelogram" method of construction, which is equivalent). The direction of the resultant is measured from the horizontal axis as usual.

$\vec{R}$ = 6.5 blocks at 12.5° N of E   2 blocks NE

5 blocks E

Figure 1.4

EXAMPLE 3

For our third example, suppose the girl walks five blocks east and then two blocks north. Figure 1.5 shows the vector diagram for this set of displacements.

Figure 1.5

EXAMPLE 4

Finally, for our fourth example, suppose the girl walks five blocks east and then two blocks west. Her final displacement will be three blocks east of the starting point as shown in Figure 1.6.

$\vec{R}$ = 3 blocks E   2 blocks W

5 blocks E

Figure 1.6

In our four examples (Figures 1.3, 1.4, 1.5, and 1.6), the two vectors of magnitudes 5 and 2 could add to any value between 7 and 3, depending on their relative orientation.

From these examples, we conclude that, as the angle between two vectors increases, the magnitude of their resultant decreases. Also, the magnitude of the resultant between two vectors is a maximum when the vectors are in the same direction (at a relative angle of 0 degrees) and a minimum when they are in the opposite direction (at a relative angle of 180 degrees). It is important to remember that the vectors must be constructed head to tail and a suitable scale chosen for the system.

Algebraic Considerations

In the above examples, the resultant between two vectors was constructed using a vector diagram. The magnitude of the resultant was the measured length of the vector drawn from the tail of the first vector to the head of the second. If we sketch such a situation, we form a triangle whose sides are related by the **law of cosines** and whose angles are related by the **law of sines**.

Consider the vector triangle in Figure 1.7 with arbitrary sides a, b, and c and corresponding angles A, B, and C.

Figure 1.7

The law of cosines states that

$$c^2 = a^2 + b^2 - 2ab\cos C$$

The law of sines states that

$$\frac{a}{\sin A} = \frac{b}{\sin B} = \frac{c}{\sin C}$$

If we use the information given in our second example above, we see that in Figure 1.8 the vectors have the following magnitudes and directions:

Figure 1.8

Using the law of cosines, we obtain the magnitude of the resultant in "blocks":

$$c = \sqrt{(2)^2 + (5)^2 - 2(2)(5)\cos 135} = 6.56$$

Using the law of sines, we find that the direction of the resultant is angle A:

$$\frac{6.56}{\sin 135} = \frac{2}{\sin A}$$

Angle A turns out to be equal to 12.45 degrees north of east.

Addition of Multiple Vectors

As long as the vectors are constructed head to tail, multiple vectors can be added in any order. If the resultant is zero, the diagram constructed will be a closed geometric figure. This can be shown in the case of three vectors forming a closed triangle, as in Figure 1.9.

$$\vec{a} + \vec{b} + \vec{c} = 0$$

Figure 1.9

Thus, if the girl walks five blocks east, two blocks north, and five blocks west, the vector diagram will look like Figure 1.10, with the resultant displacement being equal to two blocks north.

Figure 1.10

SUBTRACTION OF VECTORS

Subtraction of vectors is best understood by rethinking the subtraction as the addition of a negative vector.

$$\vec{R}_2 - \vec{R}_1 = \vec{R}_2 + (-\vec{R}_1)$$

A negative vector is simply one of the same magnitude but directed opposite (i.e., a 180-degree rotation). Note that the result of this operation is the same here as in the example that follows since the resultant has the same magnitude and direction. See Figures 1.11 and 1.12. This difference between the two vectors is sometimes called $\Delta \vec{R}$, so that $\Delta \vec{R} = \vec{R}_2 - \vec{R}_1$.

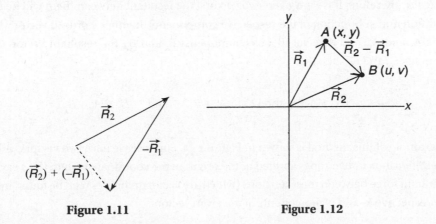

Figure 1.11   **Figure 1.12**

From our understanding of vector diagrams, we can see that $\vec{R}_1$ and $\Delta \vec{R}$ are connected head to tail, and thus that $\vec{R}_1 + \Delta \vec{R} = \vec{R}_2$!

ADDITION METHODS USING THE COMPONENTS OF VECTORS

Any single vector in space can be **resolved** into two perpendicular components in a suitably chosen coordinate system. The methods of vector resolution were discussed at the beginning of this section, but we can review them here. Given a vector (sketched in Figure 1.13) representing a displacement of 100 meters northeast (that is, making a 45-degree angle with the positive x-axis), we can observe that it is composed of an x- and a y- (a horizontal and a vertical) component that can be geometrically or algebraically determined.

Figure 1.13

Geometrically, if we were to draw the 100-meter vector to scale at the correct angle, then projecting a perpendicular line down from the head of the given vector to the x-axis would construct the two perpendicular components. Algebraically, we see that if we have a given vector $\vec{R}$, then, in the x-direction, we have

$$R_x = |\vec{R}|\cos\theta \quad \text{and} \quad R_y = |\vec{R}|\sin\theta$$

where the magnitude of $\vec{R}$ is R. The magnitude R may also be written as $|\vec{R}|$.

This method of vector resolution can be useful when adding two or more vectors. Since all vectors in the same direction add up numerically, and all vectors in opposite directions subtract numerically, therefore, if we are given two vectors, the resultant between them will be found from the addition and subtraction of the respective components. In other words, if vector $\vec{A}$ has components A_x and A_y, and if vector $\vec{B}$ has components B_x and B_y, the resultant vector $\vec{C}$ will have components C_x and C_y such that

$$\text{if } \vec{C} = \vec{A} + \vec{B}, \text{ then } C_x = A_x + B_x \quad \text{and} \quad C_y = A_y + B_y$$

An example of this method is shown in Figure 1.14. Suppose we have two vectors, $\vec{A}$ and $\vec{B}$, that represent tensions in two ropes applied at the origin of the coordinate system (let's say it's a box). The resultant force between these tensions (which are vector quantities) can be found by determining the respective x- and y-components of the given vectors.

Figure 1.14

From the diagram, we see that

$$A_x = 5 \cos 25 = 4.53 \text{ N} \quad \text{and} \quad A_y = 5 \sin 25 = 2.11 \text{ N}$$

For the second vector, we have

$$B_x = -10\cos 45 = 10\cos 135 = -7.07 \text{ N (negative since it is pointing left)}$$
$$B_y = 10\sin 45 = 10\sin 135 = 7.07 \text{ N (positive since it is pointing right)}$$

Therefore,

$$C_x = 4.53 \text{ N} - 7.07 \text{ N} = -2.54 \text{ N} \quad \text{and} \quad C_y = 2.11 \text{ N} + 7.07 \text{ N} = 9.18 \text{ N}$$

The magnitude of the resultant vector, $\vec{C}$, is given by

$$|\vec{C}| = \sqrt{(-2.54)^2 + (9.18)^2} = 9.52 \text{ N}$$

The angle that vector $\vec{C}$ makes with the positive x-axis is about 105°.

Tip-to-tail method:

SUMMARY

- Vectors are quantities that have both magnitude (size) and direction.
- Vectors can be resolved into components that represent the magnitude and direction of a vector along a particular axis.
- Scalars are quantities without direction.
- Force, displacement, and velocity are examples of vectors.
- Distance, speed, and mass are examples of scalars.
- Vectors can be "added" geometrically by the tip-to-tail method.
- The resultant of two vectors is equal to the vector obtained by "adding" them.

- If two vectors are in the same direction, their resultant is equal to the sum of their magnitudes.
- If two vectors are in the opposite direction, their resultant is equal to the difference of their magnitudes.

Problem-Solving Strategies for Vectors

When solving a vector problem, be sure to:

1. Understand the frame of reference for the situation.
2. Select an appropriate coordinate system for the situation. Cartesian systems do not necessarily have to be vertical and horizontal (a mass sliding down an inclined plane is an example of a rotated system with the "*x*-axis" parallel to the incline).
3. Pick an appropriate scale for starting your vector diagram. For example, in a problem with displacement, a scale of 1 cm = 1 m might be appropriate.
4. Recognize the given orientation of the vectors as they correspond to geographical directions (N, E, S, W). Be sure to maintain the same directions as you draw your vector triangle.
5. Connect vectors head to tail, and connect the resultant from the tail of the first vector to the head of the last vector.
6. Identify and determine the components of the vectors. In most cases, vector problems can be simplified using components.

PRACTICE EXERCISES

Multiple-Choice

1. A vector is given by its components, $A_x = 2.5$ and $A_y = 7.5$. What angle does vector $\vec{A}$ make with the positive *x*-axis?

   (A) 72°
   (B) 18°
   (C) 25°
   (D) 50°

2. Which pair of vectors could produce a resultant of 35?

   (A) 15 and 15
   (B) 20 and 20
   (C) 30 and 70
   (D) 20 and 60

3. A vector has a magnitude of 17 units and makes an angle of 20° with the positive *x*-axis. The magnitude of the horizontal component of this vector is

   (A) 16 units
   (B) 4.1 units
   (C) 5.8 units
   (D) 50 units

4. As the angle between a given vector and the horizontal axis increases from 0° to 90°, the magnitude of the vertical component of this vector

(A) decreases
(B) increases and then decreases
(C) decreases and then increases
(D) increases

5. Vector $\vec{A}$ has a magnitude of 10 units and makes an angle of 30° with the horizontal x-axis. Vector $\vec{B}$ has a magnitude of 25 units and makes an angle of 50° with the negative x-axis. What is the magnitude of the resultant of these two vectors?

(A) 20
(B) 35
(C) 15
(D) 25

6. Two concurrent vectors have magnitudes of 3 units and 8 units. The difference between these vectors is 8 units. The angle between these two vectors is

(A) 34°
(B) 56°
(C) 79°
(D) 113°

7. Which of the following sets of displacements have equal resultants when performed in the order given?

   I: 6 m east, 9 m north, 12 m west
  II: 6 m north, 9 m west, 12 m east
 III: 6 m east, 12 m west, 9 m north
 IV: 9 m north, 6 m east, 12 m west

(A) I and IV
(B) I and II
(C) I, III, and IV
(D) I, II, and IV

8. Which vector represents the direction of the two concurrent vectors shown below?

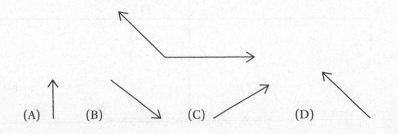

9. Three forces act concurrently on a point *P* as shown below. Which vector represents the direction of the resultant force on point *P*?

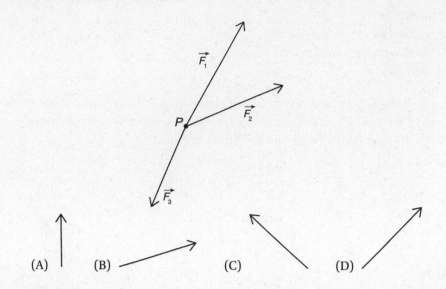

(A)   (B)   (C)   (D)

10. On a baseball field, first base is about 30 m away from home plate. A batter gets a hit and runs toward first base. She runs 3 m past the base and then runs back to stand on it. The magnitude of her final displacement from home plate is

(A) 27 m
(B) 30 m
(C) 33 m
(D) 36 m

Free-Response

1. Find the magnitude and direction of the two concurrent forces shown below using the algebraic method of components.

Note: Figure is not drawn to scale.

2. Two vectors, $\vec{A}$ and $\vec{B}$, are concurrent and attached at the tails with an angle θ between them. Given that each vector has components A_x, A_y and B_x, B_y, respectively, use the law of cosines to show that

$$\cos\theta = \frac{A_x B_x + A_y B_y}{|\vec{A}||\vec{B}|}$$

3. Give a geometric explanation for the following statement: "Three vectors that add up to zero must be coplanar (that is, they must all lie in the same plane)."

4. (a) Is it necessary to specify a coordinate system when adding two vectors?
   (b) Is it necessary to specify a coordinate system when forming the components of a vector?

5. Can a vector have zero magnitude if one of its components is nonzero?

ANSWERS EXPLAINED

Multiple-Choice Problems

1. **(A)** The angle for the vector to the positive x-axis is given by

$$\tan\theta = \frac{A_y}{A_x} = \frac{7.5}{2.5}$$

Thus $\tan\theta = 3$ and $\theta = 71.5°$ or, rounded off, $72°$.

2. **(B)** Two vectors have a maximum resultant, whose magnitude is equal to the numerical sum of the vector magnitudes, when the angle between vectors is $0°$. The resultant is a minimum at $180°$, and the magnitude is equal to the numerical difference between the magnitudes. When this logic is used, only pair $(20, 20)$ produces a maximum and minimum set that could include the given value of 35 if the angle were specified. All the others have maximum and minimum resultants that do not include 35 in their range.

3. **(A)** The formula for the horizontal component of a vector $\vec{A}$ is $A_x = |\vec{A}|\cos\theta$, where $|\vec{A}|$ is the magnitude of $\vec{A}$. When the known values are used, values $A_x = 17\cos 20 = 15.9$ units.

4. **(D)** The vertical component of a vector is proportional to the sine of the angle, which increases to a maximum at $90°$.

5. **(D)** Form the components of each vector and then add these components, remembering the sign conventions; right and up are positive values. Thus, we have $A_x = 10\cos 30 = 8.66$ and $A_y = 10\sin 30 = 5$, $B_x = -25\cos 50 = -16.06$ and $B_y = 25\sin 50 = 19.15$.

The components of the resultant are $C_x = A_x + B_x = 8.66 - 16.06$, so $C_x = -7.4$ units and $C_y = A_y + B_y = 5 + 19.15 = 24.15$. The magnitude of this resultant vector is given by

$$|\vec{C}| = \sqrt{C_x^2 + C_y^2} = \sqrt{(7.4)^2 + (24.15)^2}$$

which rounds to $|\vec{C}| = 25$ units.

6. **(C)** Use the law of cosines where $a = 3$, $b = 8$, and $c = 8$, and solve for $\cos\theta$:

$$(8)^2 = (3)^2 + (8)^2 - 2(3)(8)\cos\theta$$

Solving gives $\cos\theta = 0.1875$ and $\theta = 79°$.

7. **(C)** Vectors can be added in any order. The only requirement is that the vectors have the same magnitude and direction as they are shuffled. A look at the four sets of displacements indicates that I, III, and IV consist of the same vectors listed in different orders. These three sets will produce the same resultant.

8. **(C)** Sketch the vectors head to tail as if forming a vector triangle in construction. The resultant is drawn from the tail of the first vector to the head of the second. Choose the horizontal vector as the first one; then choice C is the general direction of the resultant.

9. **(B)** Sketch the vectors in any order, as shown below. Draw the resultant from the tail of the first vector to the head of the last.

10. **(B)** The displacement is in the direction of first base and is equal in magnitude to the straight-line distance between home plate and first base (30 m).

Free-Response Problems

1. Let $A = 10$ N force and $B = 20$ N force.

    We declare left to be negative and up to be positive.

    From the diagram, the components are:

$$A_x = 10 \text{ N} \cos 30° = 8.66 \text{ N} \quad B_x = -20 \text{ N} \cos 60° = -10 \text{ N}$$
$$A_y = 10 \text{ N} \sin 30° = 5 \text{ N} \quad\quad B_y = 20 \text{ N} \sin 60° = 17.32 \text{ N}$$

The components of the resultant are now given by the following equations. Watch your signs!

$$R_x = A_x + B_x = 8.66 \text{ N} + (-10 \text{N}) = -1.34 \text{ N}$$
$$R_y = A_y + B_y = 5 \text{ N} + 17.32 \text{ N} = 22.32 \text{ N}$$

Use the Pythagorean theorem to find the magnitude of the resultant:

$$|\vec{R}|^2 = (-1.34 \text{ N})^2 + (22.32 \text{ N})^2$$
$$|\vec{R}| = 22.36 \text{ N}$$

To find the direction, notice that the resultant will lie in the second quadrant since R_x is negative and R_y is positive:

$$\tan\theta = \frac{22.32 \text{ N}}{-1.34 \text{ N}} = -16.66$$
$$\theta = 86.6° \text{ (relative to the } x\text{-axis)}$$

2. The situation is shown below. The "third" side of the vector triangle is $\vec{A} - \vec{B}$.

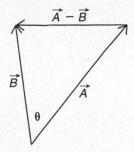

From the rules for vector subtraction, the components of $\vec{A} - \vec{B}$ are $(A_x - B_x)$ and $(A_y - B_y)$. The law of cosines states that, for the given triangle,

$$|\vec{A} - \vec{B}|^2 = |\vec{A}|^2 + |\vec{B}|^2 - 2|\vec{A}||\vec{B}|\cos\theta$$

Using our definitions of magnitudes and components, we have

$$(A_x - B_x)^2 = (A_y - B_y)^2 = A_x^2 + A_y^2 + B_x^2 + B_y^2 - 2|\vec{A}||\vec{B}|\cos\theta$$

Expanding the left side and canceling like terms on both sides, we are left with

$$-2A_x B_x - 2A_y B_y = -2|\vec{A}||\vec{B}|\cos\theta$$

Again, canceling like terms on both sides leaves us with the final expression as we solve for $\cos\theta$:

$$\cos\theta = \frac{A_x B_x + A_y B_y}{|\vec{A}||\vec{B}|}$$

3. Any two vectors can be added geometrically using the parallelogram method of construction. A parallelogram is a plane figure. The resultant of two vectors is represented by the diagonal of this parallelogram. In order for three vectors to add up to zero, the third remaining vector must be equal in magnitude, but opposite in direction, to the resultant of the remaining two vectors. Hence, all three must lie in the same plane to achieve this result.

4. (a) Using the logic of problem 3, any two vectors can be added by simply constructing a parallelogram from them. They can be in any orientation and in any plane. Hence, no coordinate system is necessary.

   (b) The components of a vector, by definition, lie along the axes of a chosen coordinate system (such as the x- or y-axis). Hence, to form these components, a coordinate system must be specified.

5. The magnitude of a vector is a number obtained by taking the square root of the sum of the squares of the magnitudes of its components. If one of the components is nonzero, the magnitude of the whole vector must, by definition, be nonzero as well.

Kinematics

2

Motion involves the change in position of an object over time. When we observe an object moving, it is always with respect to a frame of reference. Since descriptions of motion have both magnitude *and* direction, there is a vector nature to motion that must be taken into account when we want to analyze how something moves.

In this chapter, we shall confine our discussions to motion in one dimension only. In physics, the study of motion is called **kinematics**. Kinematics is a completely descriptive study of *how* something moves.

AVERAGE AND INSTANTANEOUS MOTION

Figure 2.1

TIP

Distance is a scalar quantity, while displacement is a vector quantity.

If, in Figure 2.1, we consider the actual distance traveled by the object along some arbitrary path, we are dealing with a scalar quantity. The **displacement vector $\overline{AB}$**, however, is directed along the line connecting points A and B (whether or not this is the actual route taken). Thus, when a baseball player hits a home run and runs around the bases, he or she may have traveled a distance of 360 feet (the bases are 90 feet apart), but the player's final displacement is zero (having started and ended up at the same place)!

If we are given the displacement vector of an object for a period of time Δt, we define the **average velocity**, $\vec{v}_{avg}$, to be equal to

$$\vec{v}_{avg} = \frac{\Delta \vec{x}}{\Delta x}$$

This is a vector quantity since directions are specified. Numerically, we think of the average speed as being the ratio of the total distance traveled to the total elapsed time. The units of velocity are meters per second (m/s).

If we are interested in the velocity at any instant in time, we can define the **instantaneous velocity** to be the velocity, v, as determined at any precise instant in time. In a car, the speedometer registers instantaneous speed, which can become velocity if we take into account the direction of motion. If the velocity is constant, the average and instantaneous velocities are equal, and we can write simply

$$\vec{x} = \vec{v}\Delta t$$

Note that if we were to graph time on the x-axis and displacement on the y-axis, the slope would be velocity!

Speed is a scalar quantity while velocity is a vector quantity.

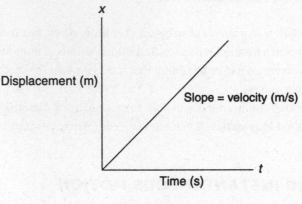

Figure 2.2

If an object is moving with a constant velocity such that its position is taken to be zero when it is first observed, a graph of the expression $\vec{x} = \vec{v}t$ would represent a direct relationship between position and time (see Figure 2.2).

Since the line is straight, the constant slope (in which $\vec{v} = \Delta \vec{x}/\Delta t$) indicates that the velocity is constant throughout the time interval. If we were to plot velocity versus time for this motion, the graph might look like Figure 2.3.

Figure 2.3

Notice that, for any time interval t, the area under the graph equals the displacement during that interval.

ACCELERATION

Velocity will change if either speed or direction changes. In any case, if the velocity is changing, we say that the object is **accelerating**. If the velocity is changing uniformly, the object has uniform acceleration. In this case, a graph of velocity versus time would look like Figure 2.4. Note that in Figure 2.3, the slope of the graph is zero. This makes sense since if the velocity is constant, acceleration is zero.

TIP

Acceleration is a vector quantity and equal to the rate of change of velocity.

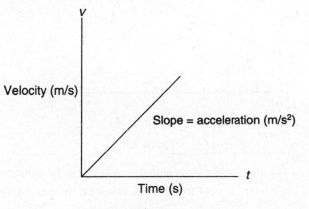

Figure 2.4

The displacement from $t = 0$ to any other time is equal to the area of the triangle formed. However, between any two intermediary times, the resulting figure is a trapezoid. If we make several measurements, the displacement versus time graph for uniformly accelerated motion is a parabola starting from the origin (if we make the initial conditions that, when $t = 0$, $x = 0$, and $v = 0$; see Figure 2.5).

Figure 2.5

The slope of the velocity versus time graph is defined to be the average acceleration in units of meters per second squared (m/s^2). We can now write, for the average acceleration (Figure 2.6),

$$\vec{a}_{avg} = \frac{\Delta v}{\Delta t}$$

Figure 2.6

If the acceleration is taken to be constant in time, our expression for average acceleration can be written in a form that allows us to calculate the instantaneous final velocity after a period of acceleration has taken place. In other words, $\Delta v = \vec{a}t$ (if we start our time interval from zero). If we define Δv to be equal to the difference between a final and an initial velocity, we can arrive at the fact that

$$\vec{v_f} = \vec{v_i} + \vec{a}t$$

If we plot velocity versus time for uniformly accelerated motion starting with a nonzero initial velocity, we get a graph that looks like Figure 2.7.

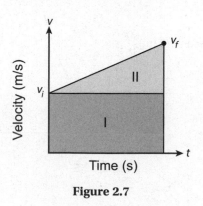

Figure 2.7

The displacement during any period of time will be equal to the total area under the graph. In this case, the total area will be the sum of two areas, one a triangle and the other a rectangle. The area of the rectangle, for some time t, is just $v_i t$. The area of the triangle is one-half the base times the height. The "base" in this case is the time period, t; the "height" is the change in velocity, Δv. Therefore, the area of the triangle is $\frac{1}{2}\Delta v t$. If we recall the definition of Δv and the fact that $v_f = v_i + at$, we obtain the following formula for the displacement during uniformly accelerated motion starting with an initial velocity:

$$x = v_i t + \frac{1}{2}at^2$$

This analysis suggests an alternative method of determining the average velocity of an object during uniformly accelerated motion. The area under the flat line halfway between v_i and v_f will be the same as in Figure 2.7.

Therefore, we can simply write

$$v = \frac{v_i + v_f}{2} = v_{avg}$$

Since the above equation relates the average velocity to the initial and final velocities (for uniformly accelerated motion), we can write our displacement formula as

$$x = \frac{v_i + v_f}{2}t = \left(v_{avg}\right)t$$

Occasionally, a problem in kinematics does not explicitly mention the time involved. For this reason, it would be nice to have a formula for velocity that does not involve the time factor. We can derive one from all the other formulas.

Now, since $v_f = v_i + at$, we can express the time as $t = (v_f - v_i)/a$. Therefore

$$x = \left(\frac{v_i + v_f}{2}\right)\left(\frac{v_f - v_i}{a}\right)$$

$$x = \left(\frac{v_f^2 - v_i^2}{2a}\right)$$

Motion can be shown graphically. Identify each region of the following graph as having negative, positive, or zero average acceleration for that interval of time.

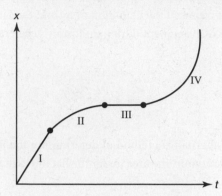

Solution

The instantaneous slopes on the given position versus time graph are the velocity values. To determine the acceleration, we must look at how those slopes are changing over the particular interval.

   I: All instantaneous slopes have the same value during this interval. There is no change in velocity: **zero acceleration**.

  II: Instantaneous slopes are decreasing (from large positive to zero) during this interval: **negative acceleration**.

 III: No change in slope equals **zero acceleration**. (Note that the velocity also happens to be zero here.)

 IV: Instantaneous slopes are increasing (from small positive values to large ones): **positive acceleration**.

A particle accelerates from rest at a uniform rate of 3 m/s^2 for a distance of 200 m. How fast is the particle going at that time? How long did it take for the particle to reach that velocity?

Solution

We use the formula

$$v_f^2 - v_i^2 = 2ax$$

Since the particle begins from rest, the initial velocity is equal to zero and

$$v_f^2 = 2(3 \text{ m/s}^2)(200 \text{ m})$$
$$v_f = 34.64 \text{ m/s}$$

To find the time, we use $v = at$

$$t = \frac{34.64 \text{ m/s}}{3 \text{ m/s}^2} = 11.55 \text{ s}$$

ACCELERATED MOTION DUE TO GRAVITY

Since velocity and acceleration are vector quantities, we need to consider the algebraic conventions accepted for dealing with various directions. For example, we usually agree to consider motion up or to the right as positive, and motion down or to the left as negative. Note that negative velocity means motion to the left (or down) and does not imply the object is slowing down. Likewise, negative acceleration means velocity is changing in the negative direction and does not imply the object is slowing down. When people say an object is decelerating, however, they do mean the object is slowing down. That can happen in two different ways. First, an object with positive velocity can slow down with a negative acceleration. Second, an object with a negative velocity can slow down with a positive acceleration. The general rule is that if the velocity and acceleration are pointed in the same direction, the object speeds up. However, if they are in opposite directions, the object slows down. (Acceleration at a right angle to velocity makes an object change direction.)

EXAMPLES:

"Slowing down"   "Speeding up"   "Turning"

Recall that acceleration is rate of change of velocity, so the direction of the acceleration can always be found from doing a quick head-to-tail graphical solution of $v_f - v_i$. For example, when undergoing circular (or any turning) motion, the acceleration can be shown to be inward:

Gravity provides a constant downward acceleration. The value of gravity on Earth is commonly represented by the symbol g to stand in for 9.8 m/s/s. Note that an object going upward will be slowed down by gravity, while a downward moving object is sped up by gravity. A sideways-moving object will turn toward the ground! An object undergoing acceleration due to gravity and gravity alone (with no other forces) is referred to as being in "free fall." This is true regardless of whether the object is moving vertically up or down. Free fall is also called "projectile motion" if the object is also moving sideways while under the influence of gravity.

REMEMBER

Students often forget that there is still gravitational acceleration even at the top of a projectile's path! The acceleration vector never changes during free fall (including during projectile motion).

Choosing up as positive and down as negative for all three kinematic vectors (displacement, velocity, and acceleration) will provide consistent results for vertical velocities and vertical displacements:

$$v_f = v_i + at \quad \text{becomes} \quad v_f = v_i - gt \quad \text{since } a = -g$$

Using Δy to represent the vertical displacement:

$$\Delta y = v_i + \frac{1}{2}at^2 \quad \text{becomes} \quad \Delta y = v_i - \frac{1}{2}gt^2$$

Note that if an object is dropped or falls, $v_i = 0$. All subsequent displacements and velocities are negative since all motion is downward in those cases. If an object is thrown upward, v_i is positive. If an object is thrown downward, v_i is negative.

SAMPLE PROBLEM

A projectile is fired vertically upward at an initial velocity of +98 m/s. How high will it rise? How long will it take the projectile to reach that height?

Solution

We use $a = -9.8$ m/s^2 (recall gravity's direction is downward) and note that $v_{fy} = 0$ at the highest point:

$$v_{fy}^2 - v_{iy}^2 = 2ay$$
$$-(98 \text{ m/s})^2 = -2(9.8 \text{ m/s}^2)y_{max}$$
$$y_{max} = 490 \text{ m}$$

To find the time, we use

$$v = at$$
$$t = \frac{-98 \text{ m/s}}{-9.8 \text{ m/s}^2} = 10 \text{ s}$$

GRAPHICAL ANALYSIS OF MOTION

We have seen that much information can be obtained if we consider the graphical analysis of motion. If complex changes in motion are taking place, visualization may provide a better understanding of the physics involved than algebra. The techniques of graphical analysis are as simple as slopes and areas. For example, we already know that, for uniformly accelerated motion, the graph of distance versus time is a parabola. Since the slope is changing, the instantaneous velocity can be approximated at a point P by finding the slope of a tangent line drawn to a given point on the curve (see Figure 2.8).

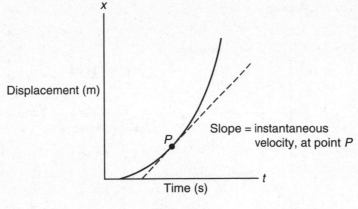

Figure 2.8

What would happen if an object accelerated from rest, maintained a constant velocity for a while, and then slowed down to a stop? Using what we know about graphs of velocity and acceleration in displacement versus time, we might represent the motion as shown in Figure 2.9.

Figure 2.9

We can apply many instances of motion to graphs. In the case of changing velocity, consider the graph of velocity versus time for an object thrown upward into the air, reaching its highest point, changing direction, and then accelerating downward. This motion has a constant downward acceleration that, at first, acts to slow the object down, but later acts to speed it up. A graph of this motion is seen in Figure 2.10.

(a)

(b)

Figure 2.10

Figure 2.10a shows the parabolic nature of $\Delta y = v_i t - \frac{1}{2}gt^2$.

Figure 2.10b shows the linear nature of $v_f = v_i - gt$.

SAMPLE PROBLEM

A ball is thrown straight up into the air and, after being in the air for 9 s, is caught by a person 5 m above the ground. To what maximum height did the ball go?

Solution

The equation

$$y = v_i t - \frac{1}{2}gt^2$$

represents the vertical position of the ball above the ground for any time t. Since the ball is thrown upward, the initial velocity is positive, while the acceleration of gravity is always directed downward (and hence is algebraically negative). Thus, we can use the above equation to find the initial velocity when $y = 5$ m and $t = 9$ s. Substituting these values, as well as the magnitude of g (9.8 m/s^2) in the equation, we get $v_i = 44.65$ m/s.

Now, to find the maximum height, we note that, when the ball rises to its maximum, its speed becomes zero since its velocity is changing direction. Since we do not know how long the ball takes to rise to its maximum height (we could determine this value if desired), we can use the formula

$$v_f^2 - v_i^2 = -2gy$$

When the final velocity equals zero, the value of y is equal to the maximum height. Using our answer for the initial velocity and the known value of g, we find that $y_{max} = 101.7$ m.

RELATIVE MOTION

All measurements are relative. Specifically, all measurements, including velocity, are made with respect to some reference point or object. Just as vectors are all relative to a specific coordinate system, all velocities are too. When we describe a car as moving at 55 mph, what we really mean is 55 mph relative to Earth's surface. The rules of vector addition provide us with a means by which to translate one relative velocity to another.

An example of this type of motion can be seen when a boat trying to cross a river or an airplane meeting a crosswind is considered. In the case of the boat, its velocity, relative to the river, is based on the properties of the engine and is measured by the speedometer on board. However, to a person on the shore, its **relative velocity** (or effective velocity) is different from what the speedometer in the boat may report. In Figure 2.11, we see such a situation with the river moving to the right at 4 meters per second and the boat moving relative to the water at 10 meters per second. If you like, we can call to the right "eastward" and up the page "northward."

Figure 2.11

By vector methods, the resultant velocity relative to the shore is given by the Pythagorean theorem. The direction is found by means of a simple sketch (Figure 2.12) connecting the vectors head to tail to preserve the proper orientation. Numerically, we can use trigonometry.

Figure 2.12

The resultant velocity is 10.5 meters per second at an angle of 22 degrees east of north.

HORIZONTALLY LAUNCHED PROJECTILES

If you roll a ball off a smooth table, you will observe that it does not fall straight down. With trial and error, you might observe that how far it falls will depend on how fast it is moving forward. Initially, however, the ball has no vertical velocity. The ability to "fall" is given by gravity, and the acceleration due to gravity is 9.8 meters per second squared downward. Since gravity acts vertically and the initial velocity (v_i) is horizontal, the two motions are simultaneous and independent. Galileo demonstrated that the trajectory characterized by constant horizontal velocity and constant vertical acceleration is a parabola.

We know that the distance fallen by a mass dropped from rest is given by the equation

$$\Delta y = v_{iy}t + \frac{1}{2}a_y t^2 = \frac{1}{2}gt^2$$

$$\text{since } v_{iy} = 0 \text{ and } a_y = g$$

Since the ball that rolled off the table is moving horizontally, with some initial constant velocity, it covers a distance, x, called the **range**:

$$\Delta x = v_{ix}t + \frac{1}{2}a_x t^2 = v_{ix}t$$

$$\text{since } a_x = 0$$

Since the time is the same for both motions, we can first solve for the time, using the x equation, and then substitute it into the y equation. In other words, y as a function of x is easily shown to be

$$\Delta y = \frac{1}{2}gt^2 = \frac{1}{2}g\left(\Delta x / v_{ix}\right)^2 = g\Delta x^2 / \left(2v_{ix}^2\right)$$

which is of course the equation of a parabola. This equation of y in terms of x is called the **trajectory** of the projectile, while the two separate equations for x and y as functions of time are called **parametric equations**. Figure 2.13 illustrates this trajectory as well as a position vector $\vec{R}$, which locates a point at any given time in space.

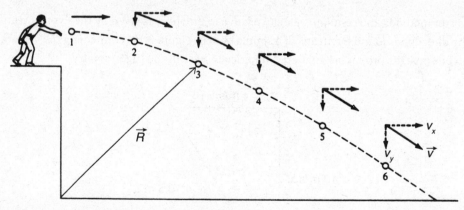

Figure 2.13

If the height from which a projectile is launched is known, the time to fall can be calculated from the equation for free fall. For example, if the height is 49 m, the time to fall is 3.16 s. If the horizontal velocity is 10 m/s, the maximum range will be 31.6 m.

$$\Delta y = \frac{1}{2}gt^2$$

$$49 = \frac{1}{2}(9.8)t^2$$

$$t = 3.2 \text{ s}$$

$$\Delta x = v_{ix}t = 10(3.2) = 32 \text{ m}$$

A projectile is launched horizontally from a height of 25 m, and it is observed to land 50 m from the base. What was the launch velocity?

Solution

We know that the vertical motion is independent of the horizontal motion. Thus, we can find the time that the projectile is in the air using:

$$d = \frac{1}{2}at^2$$

Solving for t we get

$$t = 2.26 \text{ s}$$

Now, since $x = v_x t$, we see that $v_x = \dfrac{50 \text{ m}}{2.26 \text{ s}} = 22.12 \text{ m/s}$

PROJECTILES LAUNCHED AT AN ANGLE

Suppose that a rocket on the ground is launched with some initial speed, v, at some angle θ. The vector nature of velocity allows us to immediately write the equations for the horizontal and vertical components of initial velocity:

$$v_{ix} = v\cos\theta \text{ and } v_{iy} = v\sin\theta$$

Since each motion is independent, we can consider the fact that, in the absence of friction, the horizontal velocity will be constant while the y velocity will decrease as the rocket rises. When the rocket reaches its maximum height, its vertical velocity will be zero; then gravity will accelerate the rocket back down. It will continue to move forward at a constant rate. How long will the rocket take to reach its maximum height? From the definition of acceleration and the equations earlier in this chapter, we know that this will be the time needed for gravity to decelerate the vertical velocity to zero, that is,

$$v_y = v_{iy} + at$$

$$0 = v_{iy} - gt_{up}$$

$$t_{up} = \frac{v_{iy}}{g}$$

The total time of flight will be just twice this time. Therefore, the range is the product of the initial horizontal velocity (which is constant) and the total time. In other words,

$$R = \text{range} = v_{ix}(2t_{up}) = 2v\cos\theta\, t_{up} = 2v\cos\theta(v\sin\theta/g)$$

$$= 2v^2\cos\theta\,\sin\theta/g$$

> **REMEMBER**
>
> When a projectile is launched at an angle, the vertical velocity component is equal to zero at the maximum height. Additionally, for a given launch speed, the maximum range occurs when the launch angle is equal to 45° (when launching and landing at the same height).

Since the vertical motion is independent of the horizontal motion, the changes in vertical height are given one-dimensionally as

$$\Delta y = v_{iy}t - \frac{1}{2}gt^2$$

If we want to know the maximum height achieved, we simply use the value for the time to reach the highest point. The trajectory is seen in Figure 2.14 for a baseball being hit.

Figure 2.14

As an example, if a projectile is launched with an initial velocity of 100 m/s at an angle of 30°, the maximum range will be equal to 883.7 m. To find the maximum height, we could first find the time required to reach that height, but a little algebra will give us a formula for the maximum height independent of time. The maximum height can also be expressed as

$$\Delta y = v_{iy}t + \frac{1}{2}at^2$$

$$\Delta y_{max} = v_{iy}t_{up} - \frac{1}{2}g(t_{up})^2$$

$$= \frac{1}{2}\frac{v_{iy}^2}{g}$$

Using the known numbers, we find that the maximum height reached is 127.55 m.

SAMPLE PROBLEM

A projectile is launched from the ground at a 40° angle with a velocity of 150 m/s. Calculate the maximum height of the projectile.

Solution

From trigonometry, we know that $v_{iy} = v_0 \sin\theta$. Thus,

$$v_{iy} = (150 \text{ m/s}) \sin(40°) = 96 \text{ m/s}$$

Now, at the maximum height, the vertical velocity is equal to zero, and the acceleration is g:

$$v_{fy}^2 - v_{iy}^2 = -2gy$$

$$y_{max} = \frac{v_{iy}^2}{2g} = \frac{96 \text{ m}^2/\text{s}^2}{19.6 \text{ m/s}^2} = 474 \text{ m}$$

UNIFORM CIRCULAR MOTION

Velocity is a vector. To change a velocity's direction simply requires an acceleration since change in direction is a change in the vector. An acceleration directed perpendicularly to the velocity will change the direction of the velocity in the direction of the acceleration. When the direction is the only quantity changing, as the result of a centrally directed deflecting force, the result is uniform circular motion. We consider here an object already undergoing periodic, uniform circular motion. By this description we mean that the object maintains a constant speed as it revolves around a circle of radius r, in a period of time T. The number of revolutions per second is called the **frequency**, f. This is illustrated in Figure 2.15.

Figure 2.15

The direction of the acceleration is toward the center and is called the **centripetal acceleration**. The magnitude of the centripetal acceleration is given by two formulas:

$$a_c = \frac{v^2}{r}$$

Recalling that the circumference of a circle is $2\pi r$, the speed must be:

$$V = \frac{2\pi r}{T} \text{ for uniform circular motion}$$

Substituting this expression for the tangential velocity into our formula for centripetal acceleration:

$$a_c = \frac{4\pi^2 r}{T^2}$$

Oftentimes, circular motion (or rotational motion in general) is described in terms of period or frequency of rotation rather than speed:

$$T = \frac{1}{f} \text{ (relationship between period and frequency)}$$

Sometimes, it is more convenient to describe rotational motion in terms of angular frequency (radians/sec rather than cycles/sec). Angular frequency is also the rate of change of the angular position and so is also known as the rotational (or angular) velocity. Since there are 2π radians in one complete cycle, the relationship between angular frequency (ω) and frequency is a simple one:

$$\omega = 2\pi f$$

SAMPLE PROBLEM

A 5-kg mass is undergoing uniform circular motion with a constant speed of 10 m/s in a circle of radius 2 m. Calculate the centripetal acceleration of the mass.

Solution

We use the formula

$$a_c = \frac{v^2}{r} = \frac{(10\ \text{m/s})^2}{2\ \text{m}} = 50\ \text{m/s}^2$$

SUMMARY

- Displacement, velocity, and acceleration are all vectors.
- Distance, speed, and time are scalars.
- Kinematics is a description of motion.
- Velocity is defined to be equal to the rate of change of displacement.
- Acceleration is defined to be equal to the rate of change of velocity.
- Velocity is the slope of a displacement versus time graph.
- Acceleration is the slope of a velocity versus time graph.
- The displacement can be obtained from a velocity versus time graph by taking the area under the graph.
- The acceleration due to gravity ($\vec{g}$) near the surface of Earth is equal to 9.8 m/s^2 and is directed downward. This is the only acceleration in free fall and projectile motion problems.
- The relative velocity can be found by vector addition of the individual velocity vectors.
- When a projectile is launched, in the absence of air resistance, the horizontal motion is independent of the vertical motion.
- The vertical and horizontal components of the launch velocity can be obtained using trigonometry. The regular kinematics equations can then be used for each direction.
- If an object is moving in a circle, then the object has an acceleration acting toward the center called the centripetal acceleration.

Problem-Solving Strategies for Motion Problems

Since all motion is relative, it is very important to always ask yourself the following question: "From what frame of reference am I viewing the situation?" Also, whenever you solve a physics problem, be sure to consider the assumptions being made, explicitly or implicitly, about the moving object or objects. In this way, your ability to keep track of what is relevant for your solution path will be maintained. In addition, it is suggested that you:

1. Identify all the goals and givens in the problem. Recall from the Introduction that the goals may be explicit or implicit. If a question is based on a decision or prediction, be sure to understand the algebraic requirements necessary to reach an answer.

2. Consider the meaningfulness of your solution or process. Does your answer or methodology make sense?

3. Choose a proper coordinate system and remember the sign conventions for algebraically treating vector quantities. In addition, make sure you understand the nature of the concepts being discussed (for example, whether they are vectors or scalars).

4. Use proper SI units throughout your calculations, and make sure that correct units are included in your final answer. Try dimensional analysis to see whether the answer makes sense. Perhaps it looks too large or too small because it is expressed in the wrong units.

5. Make a sketch of the situation if one is not provided.

6. If you are interpreting a graph, be sure you understand the interrelationships among all the kinematic variables discussed (slopes, areas, etc.).

7. If you are constructing a graph, label both axes completely, choose a proper scale for each axis, and draw neatly and clearly.

8. Try different problem-solving heuristics if you get stuck on a difficult problem.

9. Remember to break vectors into components first if the problem is two-dimensional.

PRACTICE EXERCISES

Multiple-Choice

1. A ball is thrown upward with an initial velocity of 20 m/s. How long will the ball take to reach its maximum height?

   (A) 19.6 s
   (B) 9.8 s
   (C) 6.3 s
   (D) 2.04 s

2. An airplane lands on a runway with a velocity of 150 m/s. How far will it travel until it stops if its rate of deceleration is constant at -3 m/s^2?

   (A) 525 m
   (B) 3,750 m
   (C) 6,235 m
   (D) 9,813 m

3. A ball is thrown downward from the top of a roof with a speed of 25 m/s. After 2 s, its velocity will be

(A) 19.6 m/s
(B) −5.4 m/s
(C) −44.6 m/s
(D) 44.6 m/s

4. A rocket is propelled upward with an acceleration of 25 m/s^2 for 5 s. After that time, the engine is shut off, and the rocket continues to move upward. The maximum height, in meters, that the rocket will reach is

(A) 900
(B) 1,000
(C) 1,100
(D) 1,200

Questions 5–7 refer to the velocity versus time graph shown below.

5. The total distance traveled by the object during the indicated 14 s is

(A) 7.5 m
(B) 25 m
(C) 62.5 m
(D) 77.5 m

6. The total displacement of the object during the 14 s indicated is

(A) 7.5 m
(B) 25 m
(C) 62.5 m
(D) 77.5 m

7. The average velocity over the whole time, in meters per second, of the object is

(A) 0
(B) 0.5
(C) 2.5
(D) 4.5

8. What is the total change in velocity for the object whose acceleration versus time graph is given below?

(A) 40 m/s

(B) −40 m/s

(C) 80 m/s

(D) −80 m/s

9. An object has an initial velocity of 15 m/s. How long must it accelerate at a constant rate of 3 m/s^2 before its average velocity is equal to twice its initial velocity?

(A) 5 s

(B) 10 s

(C) 15 s

(D) 20 s

10. A projectile is launched at an angle of 45° with a velocity of 250 m/s. If air resistance is neglected, the magnitude of the horizontal velocity of the projectile at the time it reaches maximum altitude is equal to

(A) 0 m/s

(B) 175 m/s

(C) 200 m/s

(D) 250 m/s

11. A projectile is launched horizontally with a velocity of 25 m/s from the top of a 75-m height. How many seconds will the projectile take to reach the bottom?

(A) 15.5

(B) 9.75

(C) 6.31

(D) 3.91

12. At a launch angle of 45°, the range of a launched projectile is given by

(A) $\dfrac{v_i^2}{g}$

(B) $\dfrac{2v_i^2}{g}$

(C) $\dfrac{v_i^2}{2g}$

(D) $\sqrt{\dfrac{v_i^2}{2g}}$

13. A projectile is launched at a certain angle. After 4 s, it hits the top of a building 500 m away. The height of the building is 50 m. The projectile was launched at an angle of

(A) 14°

(B) 21°

(C) 37°

(D) 76°

14. The operator of a boat wishes to cross a 5-km-wide river that is flowing to the east at 10 m/s. He wants to reach the exact point on the opposite shore 15 min after starting. With what speed and in what direction should the boat travel?

(A) 11.2 m/s at 26.6° E of N

(B) 8.66 m/s at 63.4° W of N

(C) 11.4 m/s at 60.9° W of N

(D) 8.66 m/s at 26.6° E of N

Free-Response

1. A particle has the acceleration versus time graph shown below:

If the particle begins its motion at $t = 0$, with $v = 0$ and $x = 0$, make a graph of velocity versus time.

2. A stone is dropped from a 75-m-high building. When this stone has dropped 15 m, a second stone is thrown downward with an initial velocity such that the two stones hit the ground at the same time. What was the initial velocity of the second stone?

3. A particle is moving in one dimension along the x-axis. The position of the particle is given, for any time, by the following position function:

$$x(t) = 3t^2 + 2$$

(a) Evaluate the average velocity of the particle starting at $t = 2$ s for $\Delta t = 1, 0.5, 0.2, 0.1, 0.01, 0.001$ s.

(b) Interpret the physical meaning of your results for part (a).

4. A stone is dropped from a height h and falls the last half of its distance in 4 seconds. (a) What was the total time of fall? (b) From what height was the stone dropped?

5. If the average velocity of an object is nonzero, does this mean that the instantaneous velocity of the object can never be zero? Explain.

6. A girl standing on top of a roof throws a stone straight upward into the air with a certain speed. She then throws a similar stone directly downward with the same speed. Compare the velocities of both stones when they reach the ground.

7. Explain how it is possible for an object to have zero average velocity and still have nonzero average speed.

8. A mass attached to a string is twirled overhead in a horizontal circle of radius r every 0.3 s. The path is directly 0.5 m above the ground. When released, the mass lands 2.6 m away. What was the velocity of the mass when it was released, and what was the radius of the circular path?

9. A football quarterback throws a pass to a receiver at an angle of 25 degrees to the horizontal and at an initial velocity of 25 m/s. The receiver is initially at rest 30 m from the quarterback. The instant the ball is thrown, the receiver runs at a constant velocity to catch the pass. In what direction and at what speed should he run?

10. A car is moving in a straight line with velocity v. Raindrops are falling vertically downward with a constant terminal velocity u. At what angle does the driver think the drops are hitting the car's windshield? Explain.

ANSWERS EXPLAINED

Multiple-Choice Problems

1. **(D)** Given the initial velocity of 20 m/s, we know that the ball is decelerated by gravity at a rate of 9.8 m/s^2. Therefore, we need to know how long gravity will take to decelerate the ball to zero velocity. Clearly, from the definition of acceleration, dividing 20 m/s by 9.8 m/s^2 gives the answer: 2.04 s.

2. **(B)** We do not know the time needed to stop, but we do know that the final velocity is zero. If we use the formula

$$v_f^{\,2} - v_t^{\,2} = 2ax$$

and substitute 150 m/s for the initial velocity and −3 m/s^2 for the deceleration, we get the answer: 3,750 m.

3. **(C)** The initial velocity is −25 m/s downward, so $v = -(25) - (9.8)(2) = -44.6$ m/s.

4. **(C)** The rocket accelerates from rest, so the first distance traveled in 5 s is given by

$$y_1 = \frac{1}{2}at^2$$

Substituting the values for time (5 s) and acceleration (25 m/s^2) gives $y_1 = 312.5$ m. After 5 s of accelerating, the rocket has a velocity of 125 m/s. After this time, the engine stops and the rocket is decelerated by gravity as it continues to move upward. The time to decelerate to zero is found by dividing 125 m/s by the acceleration of gravity. Thus, $125/9.8 = 12.76$ s.

The distance traveled during that time is added to the first accelerated distance. The second distance is given by

$$y_2 = 125(12.76) - 4.9(12.76)^2 = 796.6 \text{ m}$$

The total distance traveled is therefore equal to 1,108 m or, rounded off, 1,100 m.

5. **(C)** The total distance traveled by the object in 14 s is equal to the total area under the graph. Breaking the figure up into triangles and rectangles, we find that their areas add up to 62.5 m (recall that distance is a scalar).

6. **(A)** The total displacement for the object is the sum of the positive (forward) and negative (backward) areas, representing the fact that the object moves away from the origin and then back (as indicated by the velocity vs. time graph), going below the x-axis. Adding and subtracting the proper areas: $35 - 2.5 - 20 - 5 = 7.5$ m for the final displacement.

7. **(B)** Average velocity is the change in displacement for the object over the change in time (not distance). Since from question 6 we know the displacement change to be 7.5 m in 14 s, the average velocity is therefore 0.53 m/s.

8. **(A)** The change in velocity for the object is equal to the area under the graph, which is equal to $1/2 (+8)(10) = +40$ m/s.

9. **(B)** For uniformly accelerated motion, the average velocity is

$$v = \frac{v_f + v_i}{2}$$

The question requires that the average velocity be equal to twice the initial velocity; thus, if the average velocity is 30 m/s, the final velocity attained must be 45 m/s. Now the question is, how long must the object accelerate from 15 m/s to achieve a speed of 45 m/s? A change in velocity of 30 m/s at a rate of 3 m/s^2 for 10 s is implied.

10. **(B)** The horizontal component of velocity remains constant in the absence of resistive forces and is equal to $v_i \cos \theta$. Substituting the known numbers, we get $v_x = (250)(0.7) = 175$ m/s.

11. **(D)** The time to fall is given by the free-fall formula:

$$t = \sqrt{\frac{2y}{g}}$$

If we substitute the known numbers, we get 3.91 s for the time.

12. **(A)** From the formula for range, we see that, at 45°, $\sin 2\theta = 1$, and so

$$R = \frac{v_i^2}{g}$$

13. **(A)** We know that, after 4 s, the projectile has traveled horizontally 500 m. Therefore, the horizontal velocity was a constant 125 m/s and is equal to $v_i \cos \theta$. We also know that, after 4 s, the y-position of the projectile is 50 m. Thus, we can write:

$$50 = 4v_i \sin\theta - 4.9(4)^2 = 4v_i \sin\theta - 78.4$$

Therefore, $v_i \cos \theta = 125$, $v_i \sin \theta = 32.1$, and $\tan \theta = 0.2568$, so $\theta = 14.4°$.

14. **(C)** The river is flowing at 10 m/s to the right (east), and the resultant desired velocity is 5.55 m/s up (north). Therefore, the actual velocity, relative to the river, is heading W of N. By the Pythagorean theorem, the velocity of the boat must be 11.4 m/s. The angle is given by the tangent function. In the diagram below, not drawn to scale but correct for orientation, $\tan(\theta) = \dfrac{10}{5.55} = 1.8$. Therefore, $\theta = 60.9°$ W of N.

Free-Response Problems

1. The acceleration graph given in the problem indicates constant acceleration from $t = 1$ to $t = 3$. Therefore, the area represents a constant change in velocity from 0 to 9 m/s. The second region corresponds to negative acceleration, which slows the object down and turns it around. We know this since the area is -12 m/s. This brings the graph from 9 m/s to -3 m/s (a change in direction). The last area is a change of 3 m/s, which brings the velocity to 0 m/s. A graph of this motion is shown below, as desired:

2. In this problem, we know that the first stone will be dropped from a height of 75 m, and it will be in free fall because of gravity. Therefore, the time to fall is given by

$$t = \sqrt{\frac{2y}{g}} = 3.9 \text{ s}$$

Since the first stone is to fall 15 m before the second stone is dropped, we can likewise determine that $t = 1.75$ s to fall that distance. Thus, since the two stones must reach the ground at the same time, we know that the second stone must reach the ground after the first one has traveled for 3.9 s. Since the second stone is thrown when the distance fallen by the first stone is 15 m, we subtract 1.75 s from the total time of 3.9 s to get the second stone's duration of fall. That time is equal to 2.15 s. Since the second stone is thrown downward, it has a negative displacement, and so we can write

$$-75 = -v_i(2.15) - \frac{1}{2}(9.8)(2.15)^2$$

This gives us a downward initial velocity of 24.34 m/s for the second stone.

3. (a) Using the position function, we know that when $t = 2$ s, $x = 14$ m. The average velocity is given by $\Delta x / \Delta t$. For $\Delta t = 1$ s, we find that at $t = 3$ s, $x = 29$ m. Thus, $\Delta x = 15$ m and the average velocity is equal to 15 m/s. If $\Delta t = 0.5$ s, then at $t = 2.5$ s, $x = 20.75$ m; the new average velocity is equal to 13.5 m/s (6.75 m/0.5 s). We continue this procedure. For $\Delta t = 0.2$ s, the average velocity is 12.6 m/s. For $\Delta t = 0.1$ s, the average velocity is 12.3 m/s. For $\Delta t = 0.01$ s, the average velocity is 12.03 m/s. Finally, for $\Delta t = 0.001$ s, the average velocity is 12.003 m/s.

   (b) It appears that as the time interval gets smaller and smaller, the average velocity approaches 12 m/s as a limit. We can say that at t = 2 s, the "instantaneous" velocity is approximately equal to 12 m/s.

4. Let T = the total time to fall the distance h. In free fall, this is given by

$$h = \frac{1}{2}gT^2$$

Now we know that the stone falls the last half of its distance in 4 seconds. This means

$$\frac{h}{2} = \frac{1}{2}g(T-4)^2$$
$$\frac{1}{4}gT^2 = \frac{1}{2}g(T-4)^2$$
$$T^2 = 2(T-4)^2$$
$$T = \sqrt{2}(T-4)^2$$

Solving for the total time yields $T = 13.7$ seconds, and thus, $h = 915$ meters.

5. The average velocity is the ratio of the change in the displacement to the change in the time. It is possible that the object stops for a while and then continues. Thus, it is possible for a nonzero average velocity to have a zero instantaneous velocity.

6. The two stones will have the same speed when they reach the ground. As the first stone rises, gravity decelerates it until it stops, and then it begins to fall back down. When it passes its starting point, it has the same speed but in the opposite direction. This is the same starting speed as that of the second stone. Both stones are then accelerated through the same displacement, giving them both the same final velocity.

7. The average velocity is the total displacement divided by the total time and is a vector quantity. The average speed is equal to the total distance divided by the total time and is a scalar quantity. If an object is undergoing periodic motion, it can return to its starting point and have zero displacement in one period. This gives it zero average velocity. However, since it traveled a distance, it has a nonzero average speed.

8. The mass is undergoing uniform circular motion. Therefore, it has a constant speed. The velocity when it is released can be determined from the height of the mass and the horizontal range.

    We know it falls vertically 0.5 m in 0.32 s. During this time, it moves horizontally 2.6 m. Thus, the magnitude of the horizontal initial velocity was

    $$v = \frac{2.6 \text{ m}}{0.32 \text{ s}} = 8.125 \text{ m/s}$$

    In circular motion, the constant velocity found above is equal to the ratio of the circumference of the circular path and the period:

    $$v = \frac{2\pi r}{T}$$

    Substituting for the velocity and the period of 0.3 s, we find that $r = 0.388$ m.

9. The first quantity we can calculate is the football's theoretical range:

    $$R = \frac{v_i^2 \sin 2\theta}{g} = \frac{(25)^2 (\sin 50)}{g} = 48.85 \text{ m}$$

    The receiver needs to travel 18.85 m away from the quarterback to catch the ball. To determine how fast the receiver must run, we need to know how long it takes the ball to travel 48.85 m horizontally. This is equal to the range divided by the horizontal velocity:

    $$t = \frac{48.85 \text{ m}}{v_i \cos 25} = 2.16 \text{ s}$$

    Therefore, to travel 18.85 m in 2.16 s, the receiver must run at 8.73 m/s.

10. From the diagram below, we can see that, to the driver, $\tan \theta = u/v$.

Forces and Newton's Laws of Motion

3

KEY CONCEPTS
→ **FORCES**
→ **NEWTON'S LAWS OF MOTION**
→ **STATIC APPLICATIONS OF NEWTON'S LAWS**
→ **FREE-BODY DIAGRAMS**
→ **DYNAMIC APPLICATIONS OF NEWTON'S LAWS**
→ **CENTRAL FORCES**
→ **FRICTION**

From our discussion in Chapter 1 of frames of reference, you should be able to convince yourself that, if two objects have the same velocity, the relative velocity between them is zero, and therefore one object looks as though it is at rest with respect to the other. In fact, it would be impossible to decide whether or not such an **"inertial"** observer was moving! Therefore, accepting this fact, we state that an object that appears to be at rest in the Earth frame of reference will simply be stated to be "at rest" relative to us (the observers).

FORCES

An object does not change its velocity (accelerate) on its own. Rather, an interaction is required between that object and another object. This interaction can be direct as in the case of **contact force**, or it may take place via an action at a distance. Gravitational, electrical, and magnetic forces are all examples of forces that act at a distance (no direct contact is required). Friction and normal forces are examples of contact forces. If there is a net unbalanced force (the forces on an object do not add up to zero), then there will be an acceleration. This acceleration can be a change in either speed or direction of motion of the object, or it can be both. If there is no change in speed or direction, the object is not accelerating and the net force must be zero.

Unless an object is in deep space, far from the influence of any other objects, objects experience a net force of zero because two or more forces are in opposition to each other, canceling out each other. These objects experiencing a net force of zero are in a force equilibrium. If the object is at rest, it is in **static equilibrium**. Note that an object can be in motion while experiencing a net force of zero; it may be experiencing a constant velocity.

The following is a complete list of all common forces used on the AP Physics 1 exam.

TIP

Forces are vector quantities that represent a push or a pull on an object.

Types of Forces

Force	When Force Is Present	Rule for Direction
Weight/gravity	Almost always (the object could be *very* far from any planet)	Drawn straight toward the center of Earth
Normal	When an object is in contact with another object	Drawn straight away from the surface (perpendicular)
Tension	When an object is being pulled by a rope, string, or chain	Drawn in the direction the rope is attached
Push or pull	When a person or an active agent is interacting with the object	Made clear by the situation
Spring/restorative	When a spring or an elastic material is compressed or extended	Opposite the extension or compression
Friction	Surface friction: When two surfaces are sliding (kinetic) or trying to slide (static) past each other Fluid friction: When an object is moving through a gas or liquid	Opposes the motion (velocity vector) or the likely motion if stationary
Electric	Whenever two or more net charges are present	Opposite charges attract (center toward center) and similar charges repel (directly away from each other)

Of these forces, weight (or the force due to gravity when on the surface of a planet) is the most ubiquitous and has its own formula:

$$F_g = mg$$

In this equation, g is the strength of the gravity field at the location of the mass (9.8 N/kg at Earth's surface). Note that weight is a vector force and depends on an object's location in the universe. In contrast, mass is a scalar measure of an object's inertia, which belongs to the object independent of location. This g identified here as the gravitational field strength is the same g used in kinematics as the acceleration due to gravity in free fall and projectile motion.

When generating a list of everyday forces such as the one above, it is not always clear what to call a particular force. For example, if you push something with your hand, it could be called a normal force between two surfaces or it could be called simply a "push." These so-called contact forces (that is, all the forces in the table above except for gravity and electricity) are actually not fundamental in the sense that we are naming some overall effect that derives from many microscopic fundamental interactions. All forces on this list, except for gravity, are actually electromagnetic in nature. They are electromagnetic in the sense that they can ultimately be explained in terms of charges interacting with other charges. In fact, all of the forces you experience in your daily life are either gravitational or electromagnetic. Rounding out the list of fundamental forces are the strong and weak nuclear forces.

If all the forces acting on an object produce no net change, the object is in a state of **equilibrium**. If the object is moving relative to us, we say that it is in a state of **dynamic equilibrium**. If the object is at rest relative to us, we say that it is in a state of **static equilibrium**. Figure 3.1 illustrates these ideas.

Figure 3.1

NEWTON'S LAWS OF MOTION

In 1687, at the urging of his friend Edmund Halley, Isaac Newton published his greatest work. It was titled the *Mathematical Principles of Natural Philosophy* but is more widely known by a shortened version, *Principia*. In this book, Newton revolutionized the rational study of mechanics by the introduction of mathematical principles that all of nature was considered to obey. Using his newly developed ideas, Newton set out to explain the observations and analyses of Galileo Galilei and Johannes Kepler.

The ability of an object to resist a change in its state of motion is called **inertia**. This concept is the key to Newton's first law of motion:

> *Every body continues in its state of rest, or of uniform motion in a straight line, unless it is compelled to change that state by forces acting on it.*

In other words, an object at rest will tend to stay at rest, and an object in motion will tend to stay in motion, unless acted upon by an external force. By "rest," we of course mean the observed state of rest in a particular frame of reference. As stated above, the concept of "inertia" is taken to mean the ability of an object to resist a force attempting to change its state of motion. As we will subsequently see, this concept is covered under the new concept of **mass** (a scalar, as opposed to **weight**, a vector force).

If a mass has an unbalanced force incident upon it, the velocity of the mass is observed to change. The magnitude of this velocity change depends inversely on the amount of mass. In other words, a force directed along the direction of motion will cause a smaller mass to accelerate more than a larger mass. Newton's second law of motion expresses these observations as follows:

> *The change of motion is proportional to the net applied force, and is in the direction which that force acts.*

The acceleration produced by the force is in the same direction as the force. The second law is sometimes expressed as: $\vec{F} = m\vec{a}$.

However, to preserve the vector nature of the forces, and the fact that by "force" we mean "net force," we write the second law as:

$$\vec{F}_{net} = \sum \vec{F} = m\vec{a}$$

The units of force are **newtons**; 1 newton (N) is defined as the force needed to give a 1-kilogram mass an acceleration of 1 meter per second squared. Thus 1 newton equals 1 kilogram-meter per second squared.

TIP

Make sure you know all of Newton's laws. On an AP free-response problem, you should make a declaration like $\sum \vec{F} = 0$ (for equilibrium) or $\sum \vec{F} = m\vec{a}$ (for accelerated systems).

The normal force is NOT the reaction force to gravity.

Newton's third law of motion is crucial for understanding the conservation laws we will discuss later. It stresses the fact that forces are the result of mutual interactions and are thus produced in pairs. Sometimes referred to as the "action-reaction" law, the third law can be stated as follows (Figure 3.2):

For every force, there is an equal but opposite force on the object causing the original force.

Objects A and B are in contact or colliding.

They exert equal and opposite forces on each other:

Force of B on A   Force of A on B

(a)

Objects C and D are attracted to each other.

Force of D on C   Force of C on D

(b)

Figure 3.2

Remember these tips for Newton's third law action-reaction pairs.

- There are no unpaired forces in the universe. Recall that all forces arise from an interaction between two objects.
- Action and reaction forces are applied to two different objects. The pair of forces do not cancel each other for this reason.
- One force does not cause the other. Which force is the action and which is the reaction is a matter of perspective. They happen simultaneously.
- Action-reaction force pairs are the same type of force. In other words, the reaction to a normal force must be another normal force, in the opposite direction, on the other surface. For example, a normal force would never be the reaction force to a gravitational force.

SAMPLE PROBLEM

A 2-kg mass is accelerated by a net force of 20 N. What is the acceleration of the mass?

Solution

We use Newton's second law:

$$\vec{F}_{net} = m\vec{a}$$

Thus,

$$a = \frac{20 \text{ N}}{2 \text{ kg}} = 10 \text{ N/kg} = 10 \text{ m/s}^2$$

How much does a 2-kg mass weigh on the surface of Earth?

Solution

We use the formula for weight:

$$F_g = mg$$

$$F_g = (2 \text{ kg})(9.8 \text{ m/s}^2) = 19.6 \text{ N}$$

Notice in both problems that the units for acceleration can be written as both N/kg or m/s^2. They are equivalent. We can think of the acceleration due to gravity in terms of the force per unit mass (which is referred to as a field strength).

Spencer is discussing orbital motion with Oliver. Oliver claims that Earth must be exerting a stronger gravitational force back on the Moon since the Moon orbits Earth and not the other way around. Spencer points out that this cannot be the case because the two gravitational forces are an action-reaction pair and thus must be equal in magnitude. Oliver maintains that cannot be true since Earth is much bigger than the Moon.

Which student is making the better argument? Explain your answer.

Solution

Spencer is correct. The gravitational pull of Earth on the Moon and that of the Moon on Earth is an action-reaction pair. These forces are the same interaction, and one gravitational attraction could not exist without the other. The different sizes or masses of the two objects do not create different forces for each other—the Moon and Earth each experience the same force. What would change, however, is the acceleration each feels in response to the force. The Moon, with its lower inertia, experiences a much larger change in its motion than Earth.

STATIC APPLICATIONS OF NEWTON'S LAWS

If we look more closely at Newton's second law of motion, we see an interesting implication. If the net force acting on an object is zero, the acceleration of the object will likewise be zero. Notice, however, that kinematically, zero acceleration does not imply zero motion! It simply indicates that the velocity of the object is not changing. If the object is in a state of rest and remains at rest (because of zero net forces), the object is in static equilibrium. Some interesting problems in engineering deal with the static stability of structures. Let's look at a simple example.

Place this book on a table. You will observe that it is not moving relative to you. Its state of rest is provided by the zero net force between the downward force of gravity and the upward reactive

> **THINK ABOUT IT**
>
> You are pulling up on Earth with the same force that Earth is pulling down on you (i.e., your weight)!

force of the table (pushing on the floor, which in turn pushes up on the table, etc.). Figure 3.3 shows this setup. The upward reactive force of the table, sometimes called the **normal force**, always acts perpendicular to the surface.

Figure 3.3

> **TIP**
>
> **The normal force is not always equal to an object's weight. It must be found via Newton's second law.**

Remember: The second law is a vector equation, and so we must treat the sum of all vector forces in each direction separately! In this case, we have:

$$\sum F_y = N - F_g = 0$$

In this example, the normal force is equal to the object's weight.

There are no forces in the x-directions to be "analyzed."

A slightly more complex problem would concern a mass suspended by two strings. The tensions in the strings support the mass and are thus vector forces acting at angles. In the problem illustrated in Figure 3.4, a 10-kilogram mass is suspended by two strings, making angles of 30 and 60 degrees, respectively, to the horizontal. The question is, "What are the tensions in the two strings?"

Figure 3.4

FREE-BODY DIAGRAMS

A useful heuristic for solving these types of problems is to construct what is called a "free-body diagram." In such a diagram, you "free" the body of its realistic constraints and redraw it, indicating the directions of all applied forces.

You must understand that the College Board has very specific expectations for free-body diagrams. The central object experiencing the forces should be represented by a big round dot. All forces experienced by the object should be represented by one arrow coming out from that dot with no stacking (forces in the same direction should be next to each other) and no components indicated. The net force, acceleration, and velocity vector are *not* part of the free-body diagram. For example, free-body diagrams for Figures 3.3 and 3.4 would be

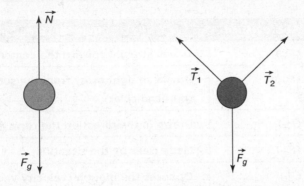

You then use that diagram to set up your static equations for Newton's second law in both the *x*- and *y*-directions. The result is a system of two equations and two unknowns. The two representations of the free-body diagram in Figure 3.5 are useful in identifying the horizontal and vertical components of the forces involved for Figure 3.4. However, these pictures of the components should not be mistaken for an actual free-body diagram.

Figure 3.5

We can now set up our equations for Newton's second law, using the techniques of vector analysis reviewed in Chapter 1. We thus determine the *x*- and *y*-components of the tensions, which are given by $T_{1x} = -T_1 \cos 30$ and $T_{2x} = T_2 \cos 60$. The negative sign is used since T_{1x} is to the left. Thus, in the *x*-direction we write $\sum F_x = ma_x = 0$ since there is no horizontal acceleration.

$$\sum F_x = 0 = T_2 \cos 60 - T_1 \cos 30 = 0.5 T_2 - 0.866 T_1$$

Likewise, for the *y*-direction, we see that the 10-kg mass weighs 98 N. The downward pull of gravity is compensated for by the upward pull of the two vertical components of the tensions. In other words, we can write for the *y*-direction $\sum F_y = ma_y = 0$ since there is no vertical acceleration.

$$\sum F_y = 0 = T_1 \sin 30 + T_2 \sin 60 - F_g = 0.5 T_1 + 0.866 T_2 - 98$$

The two simultaneous equations can then be solved for T_1 and T_2. Performing the algebra, we find that $T_1 = 49$ N and $T_2 = 85$ N.

HOW TO DRAW EFFECTIVE FREE-BODY DIAGRAMS

The forces, their standard abbreviations, and the rule for their direction are as follows:

Force	Abbreviation	Rule for Direction
Weight	(F_g)	Drawn straight toward the center of Earth
Normal	(F_N)	Drawn straight away from the surface (perpendicular)
Tension	(F_T)	Drawn in the direction the rope is attached
Push or Pull	(F_P)	Made clear by the situation
Friction	(F_f)	Opposes the motion (velocity vector) or the likely motion if stationary

1. Draw a force diagram/free-body diagram
2. Break every force into its x- and y-components
3. Add up all the x-components and set equal to ma_x
4. Add up all the y-components and set equal to ma_y
5. Decide if some (or both) components of acceleration are zero
6. Solve for the unknowns

Example: An object has three forces pulling on it in the directions shown.

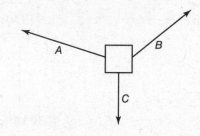

Choose your x- and y-axis based on the motion of the object. (One axis should be in the direction of acceleration!) Break vectors A and B into components using trigonometry. Then add up the x-components separately from the y-components.

In the chart a sign is added in front of each component to emphasize the importance of dealing with direction.

	A	B	C	$F_{net} = ma$
X-components	$-A_x$	$+B_x$	0	$B_x - A_x = ma_x$
Y-components	$+A_y$	$+B_y$	$-C$	$B_y + A_y - C = ma_y$

Use the final column as the starting point for solving the particular problem. At least one of the components of acceleration will be zero in almost all cases. Since we only have two equations, there can only be two unknowns!

Another static situation occurs when a mass is hung from an elastic spring. It is observed that, when the mass is attached vertically to a spring that has a certain natural length, the amount of stretching, or elongation, is directly proportional to the applied weight. This relationship, known as **Hooke's law**, supplies a technique for measuring static forces. Mathematically, Hooke's law is given as

$$\vec{F} = -k\vec{x}$$

where k is the spring constant in units of newtons per meter (N/m), and $\bar{x}$ is the elongation beyond the natural length. The negative is used to indicate that the applied force is restorative so that, if allowed, the spring will accelerate back in the opposite direction. As a static situation, a given spring can be calibrated for known weights or masses, and thus used as a "scale" for indicating weight or other applied forces.

DYNAMIC APPLICATIONS OF NEWTON'S LAWS

If the net force acting on an object is not zero, Newton's second law implies the existence of an acceleration in the direction of the net force. Therefore, if a 10-kilogram mass is acted on by a 10-newton force from rest, the acceleration will be 1 meter per second squared. Heuristically, we can construct a free-body diagram for the system to analyze all forces acting in all directions and then apply the second law of motion. We must be careful, however, to choose an appropriate frame of reference and coordinate system. For example, if the mass is sliding down a frictionless incline, a natural coordinate system to use is one that is rotated in such a way that the x-axis is parallel to the incline (see Figure 3.6).

Free-body diagram

Figure 3.6

Therefore, in order to resolve the force of gravity into components parallel and perpendicular to the incline, we must first identify the relevant angle in the geometry. This is outlined in Figure 3.6, and we can see that the component of weight along the incline is given by $-mg\sin\theta$, where θ is the angle of the incline. The magnitude of the normal force, perpendicular to the incline, is therefore given by $|\vec{N}| = |mg\cos\theta|$ since it is canceling the other component of weight.

If, for example, the mass was 10 kg, the weight would be 98 N. If $\theta = 30°$, we would write for the x-forces:

$$\sum F_x = ma = -mg\sin\theta = -98\sin 30 = -49 \text{ N}$$

giving us an acceleration of 4.01 m/s^2 down the incline. Since there is no acceleration in the y-direction, we would have:

$$\sum F_y = 0 = N - mg\cos\theta = N - 98\cos 30 = N - 84.9$$

Thus the normal force is equal to 85 N.

CENTRAL FORCES

Consider a point mass moving around a circle, supported by a string making a 45-degree angle to a vertical post (the so-called **conical pendulum**). Let's analyze this situation, which is shown in Figure 3.7.

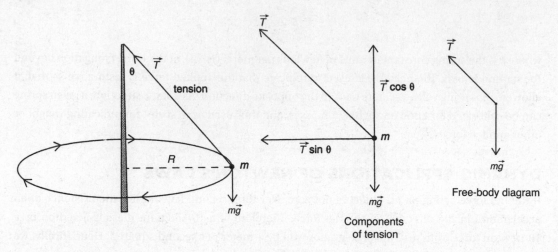

Figure 3.7

In this case, the magnitude of the weight, mg, is balanced by the upward component of the tension in the string, given by $T\cos\theta$. The inward component of tension, $T\sin\theta$, is responsible for providing a **centripetal force**. Note that the centripetal force is *not* an additional force. It is just a description of existing force(s).

If we specify that the radius is 1.5 m and the velocity of the mass is unknown, we see that, vertically, $T\cos 45 = mg$. This implies that, if $mg = 98$ N, then $T = 138.6$ N.

Horizontally, we see that

$$\sum F_x = ma_x$$
$$T\sin 45 = ma_c$$

where a_c denotes a centripetal acceleration.

$$a_c = \frac{v^2}{r}$$

Therefore, the general expression for centripetal force is given by

$$F_c = \frac{mv^2}{r}$$

using the logic of $\vec{F} = m\vec{a}$. Using our known information, we solve for velocity and find that $v = 3.8$ m/s. It is important to remember that the components of forces must be resolved along the principal axes of the chosen coordinate system.

FRICTION

Friction is a contact force between two surfaces that is responsible for opposing sliding motion. Even the smoothest surfaces are microscopically rough, with peaks and valleys like a mountain range. If a spring balance is attached to a mass and then pulled, the reading of the force scale before the mass first begins to move provides a measure of the static friction. Once the mass is moving, if we maintain a steady enough force, the velocity of movement will be constant. Thus, the acceleration will be zero, indicating that the net force is zero. The reading of the scale will then measure the kinetic friction. This reading will generally be less than the starting reading.

Observations show that the frictional force is directly related to the applied load pushing the mass into the surface. From our knowledge of forces, the normal force is measuring this "push" into the surface. The proportionality constant linking the normal force with the frictional force is called the coefficient of friction and is symbolized by μ. There are two coefficients of friction, one for static friction (μ_s) and the other for kinetic friction (μ_k). This linear relationship is often written as

$$f_k = \mu_k N$$

and the analysis of a situation involves identifying the normal force.

For example, in Figure 3.8 a mass is being pulled along a horizontal surface by a string, making an angle θ.

Figure 3.8

Newton's second law in the vertical direction gives us:

$$\Sigma F_y = ma_y = 0 \quad \text{(no vertical acceleration)}$$

$$N + T\sin\theta - mg = 0$$

Solving for N:

$$N = mg - T\sin\theta$$

Note that the normal force is less than the weight of the mass by precisely the upward component of the tension. Newton's second law in the horizontal direction gives us:

$$\Sigma F_x = ma_x$$

$$T\cos\theta - f = ma_x$$

Using our model for friction and substituting our expression for the normal force found above:

$$f = \mu_k N = \mu_k(mg - T\sin\theta)$$

With this expression for friction, we can substitute into our horizontal force equation to find the general solution:

$$T\cos\theta - \mu_k(mg - T\sin\theta) = ma_x$$

These types of problems will usually either give all the parameters (T, m, θ, μ_k) and ask for the resulting acceleration or will state the mass is being pulled with constant velocity $(a_x = 0)$, allowing you to solve for a different unknown (typically μ_k).

SAMPLE PROBLEM

A 2-kg mass is held at rest at the top of an incline that makes a 40° angle with the ground. The incline is 1.5 m long and the coefficient of kinetic friction $\mu_k = 0.2$. When the mass is released, it slides down the incline with uniform acceleration.

(a) Calculate the acceleration of the mass.
(b) Calculate the speed of the mass when it reaches the bottom of the incline.

Solution

(a)

From the geometry, we see that $F_\parallel = mgF\sin\theta$ and $F_\perp = mg\cos\theta$.

We also know that $F_f = \mu_k F_N$.

Taking y to be perpendicular to the plane and x to be down along the plane:

$\Sigma F_y = ma_y$

$\theta = 0$ (since $a_y = 0$)

$N = mg\cos\theta$

$\Sigma F_x = ma_x$

$$mg\sin\theta - F_f = ma_x$$

$$mg\sin\theta - \mu_k N = ma_x$$

$mg\sin\theta - \mu_k mg\cos\theta = ma_x$ (substituting our expression for N above)

$a_x = g(\sin\theta - \mu_k\cos\theta)$ (canceling mass and factoring out g)

$a_x = 4.8$ m/s^2 (substituting all our given values)

(b) $v_f^2 = v_i^2 + 2ad$

Since the object starts from rest, $v_i = 0$.

$$v_f^2 = 2ad = 2(4.8 \text{ m/s}^2)(1.5 \text{ m})$$

$$v_f = 3.8 \text{ m/s}$$

SUMMARY

- Forces are pushes and pulls that can be represented by vectors.
- Inertia is the tendency of a mass to resist a force changing its state of motion in a given frame of reference.
- The normal force is a force acting perpendicular to, and away from, a surface.
- Friction is a force that opposes relative motion between two objects.
- Newton's three laws help us to understand dynamics, the actions of forces on masses:
  i. In the absence of an external net force, an object maintains constant velocity (law of inertia).
  ii. If a net force is acting on a mass, then the acceleration is directly proportional to the net force ($\vec{F}_{net} = m\vec{a}$).
  iii. For every action force, there is an equal but opposite reaction force.
- Weight is a force caused by the pull of Earth's gravity on a mass and is directed downward.
- An inertial frame of reference is a frame of reference moving at a constant velocity.
- Free-body diagrams assist in the analysis of forces by identifying and labeling all forces acting on a mass freed from the confines of the illustrated situation.
- In uniform circular motion, the net force is directed inward toward the center and is referred to as the centripetal force. A centrifugal force is a fictitious or pseudo force identified in an accelerated frame of reference attached to a mass undergoing circular motion.

Be careful about the changing roles of the forces in vertical circular motion! Radially inward forces are positive and outward ones are negative. Compare the expressions for net force (centripetal force) at the top and bottom of a roller coaster loop.

$$F_{net} = F_c = N - F_g \qquad\qquad F_{net} = F_c = F_g + N$$

PRACTICE EXERCISES

Multiple-Choice

1. In the situation shown below, what is the tension in string 1?

    (A) 69.3 N
    (B) 98 N
    (C) 138.6 N
    (D) 147.6 N

2. Two masses, M and m, are hung over a massless, frictionless pulley as shown below. If $M > m$, what is the downward acceleration of mass M?

(A) g

(B) $\dfrac{(M-m)g}{M+m}$

(C) $\left(\dfrac{M}{m}\right)g$

(D) $\dfrac{Mmg}{M+m}$

3. A 0.25-kg mass is attached to a string and swung in a vertical circle whose radius is 0.75 m. At the bottom of the circle, the mass is observed to have a speed of 10 m/s. What is the magnitude of the tension in the string at that point?

(A) 2.45 N

(B) 5.78 N

(C) 22.6 N

(D) 35.7 N

4. A car and driver have a combined mass of 1,000 kg. The car passes over the top of a hill that has a radius of curvature equal to 10 m. The speed of the car at that instant is 5 m/s. What is the force of the hill on the car as it passes over the top?

(A) 7,300 N up

(B) 7,300 N down

(C) 12,300 N up

(D) 12,300 N down

5. A hockey puck with a mass of 0.3 kg is sliding along ice that can be considered frictionless. The puck's velocity is 20 m/s. The puck now crosses over onto a floor that has a coefficient of kinetic friction equal to 0.35. How far will the puck travel across the floor before it stops?

(A) 3 m

(B) 87 m

(C) 48 m

(D) 58 m

6. A spring with a stiffness constant $k = 50$ N/m has a natural length of 0.45 m. It is attached to the top of an incline that makes a 30° angle with the horizontal. The incline is 2.4 m long.

A mass of 2 kg is attached to the spring, causing it to be stretched down the incline. How far down the incline does the end of the spring rest?

(A) 0.196 m

(B) 0.45 m

(C) 0.646 m

(D) 0.835 m

7. A 20-N force is pushing two blocks horizontally along a frictionless floor as shown below.

What is the force that the 8-kg mass exerts on the 2-kg mass?

(A) 4 N

(B) 8 N

(C) 16 N

(D) 20 N

8. A force of 20 N acts horizontally on a mass of 10 kg being pushed along a frictionless incline that makes a 30° angle with the horizontal, as shown below.

The magnitude of the acceleration of the mass up the incline is equal to

(A) $1.9 \, \text{m/s}^2$

(B) $2.2 \, \text{m/s}^2$

(C) $3.17 \, \text{m/s}^2$

(D) $3.87 \, \text{m/s}^2$

9. According to the diagram below, what is the tension in the connecting string if the table is frictionless?

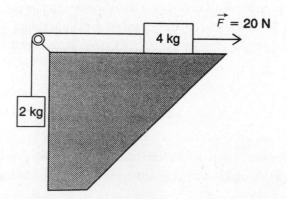

(A) 6.4 N

(B) 13 N

(C) 19.7 N

(D) 25 N

10. A mass, M, is released from rest on an incline that makes a 42° angle with the horizontal. In 3 s, the mass is observed to have gone a distance of 3 m. What is the coefficient of kinetic friction between the mass and the surface of the incline?

(A) 0.8

(B) 0.7

(C) 0.6

(D) 0.5

Free-Response

1. A 3-kg block is placed on top of a 7-kg block as shown below.

The coefficient of kinetic friction between the 7-kg block and the surface is 0.35. A horizontal force F acts on the 7-kg block.

(a) Draw a free-body diagram for each block.

(b) Calculate the magnitude of the applied force F necessary to maintain an acceleration of 5 m/s².

(c) Find the minimum coefficient of static friction necessary to prevent the 3-kg block from slipping.

2. A curved road is banked at an angle θ, such that friction is not necessary for a car to stay on the road. A 2,500-kg car is traveling at a speed of 25 m/s, and the road has a radius of curvature equal to 40 m.

(a) Draw a free-body diagram for the situation described above.

(b) Find angle θ.

(c) Calculate the magnitude of the force that the road exerts on the car.

3. The "rotor" is an amusement park ride that can be modeled as a rotating cylinder, with radius r. A person inside the rotor is held motionless against the sides of the ride as it rotates with a certain velocity. The coefficient of static friction between a person and the sides is μ.

(a) Derive a formula for the period of rotation T, in terms of r, g, and μ.

(b) If $r = 5$ m and $\mu = 0.5$, calculate the value of the period T in seconds.

(c) Using the answer to part (b), calculate the angular velocity ω in radians per second.

4. How does the rotation of Earth affect the apparent weight of a 1-kg mass at the equator?

5. If forces occur in action-reaction pairs that are equal and opposite, how is it possible for any one force to cause an object to move?

ANSWERS EXPLAINED

Multiple-Choice Problems

1. **(A)** We need to apply the second law for static equilibrium. This means that we must resolve the tensions into their x- and y-components. For T_1, we have $T_1 \cos 45$ and $T_1 \sin 45$. In equilibrium, the sum of all the x forces must equal zero. This means that $T_1 \cos 45 = T_2$ (since T_2 is entirely horizontal). The y-component of T_1 must balance the weight $= mg = 49$ N. Thus, $T_1 \sin 45 = 49$ N and $T_1 = 69.3$ N.

2. **(B)** The free-body diagrams for both masses, M and m, look like this:

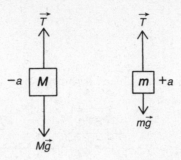

The large mass is accelerating downward, while the small mass is accelerating upward. The tension in the string is directed upward, while gravity, given by the weight, is directed downward. Using the second law for accelerated motion, we must show that for each mass, $\Sigma F_y = ma$. Thus we have

$$T - Mg = -Ma \quad \text{and} \quad T - mg = ma$$

Eliminating the tension T and solving for a gives us $(M - m)g/(M + m)$.

3. **(D)** A sketch of the situation is shown below:

At the lowest point, the downward force of gravity is opposed by the upward tension in the string. Note that doing a quick tip-to-tail graphical solution of the force vectors must result in an upward (inward) net force. Thus, we can write

$$T - mg = \frac{mv^2}{r}$$

Substituting in the given values and solving for T gives us 35.7 N.

4. **(A)** As the car goes over the hill, the force that the hill exerts on the car is the normal force, $\vec{N}$. This is opposed, however, by the downward force of gravity. The combination produces (as in question 3) the centripetal force $\vec{F_c}$. Using our circular motion convention of inward forces being positive, we can write

$$mg - N = \frac{mv^2}{r}$$

A sketch is shown below. Note that doing a quick tip-to-tail graphical solution of the force vectors must result in a downward (inward) net force. Substituting in the given values and solving for N gives 7,300 N upward.

5. **(D)** With horizontal motion, the normal force is equal to the force of gravity. When the puck is moving with constant velocity, no net force is acting on it. Thus, when friction acts to slow the puck down, the friction is the only net force. Therefore, we write

$$f = \mu N = \mu mg = -ma$$

(since the puck is decelerating). Substituting the known numbers gives a deceleration of -3.43 m/s^2. Now, the final velocity will be zero; and since the time to stop is unknown, we use the following kinematic expression to solve for the stopping distance:

$$-v_i^2 = 2ax$$

Substituting the known numbers, we obtain approximately 58 m as the distance the puck will travel before stopping.

6. **(C)** A sketch of the situation is given below. The spring constant is $k = 50$ N/m.

According to Hooke's law, $F_s = kx$, where x is the elongation in excess of the spring's natural length (here 0.45 m). The force, in this case, is provided by the component of weight parallel to the incline, which is given by $mg\sin\theta$. Substituting the known numbers gives us an elongation of $x = 0.196$ m. Since this must be in excess of the spring's natural length, the answer is 0.646 m.

7. **(C)** The 20-N force is pushing on a total mass of 10 kg. Thus, using $\vec{F} = m\vec{a}$, we have the acceleration of both blocks equal to 2 m/s^2. We draw a free-body diagram for the 2-kg mass as shown below.

Let P represent the force that the 8-kg mass exerts on the 2-kg mass. Writing the second law of motion, we get $\vec{F} - P = m\vec{a}$. To find P, we substitute the known numbers; $20 - P = (2)(2) = 4$. Thus, $P = 16$ N.

8. **(C)** From the given diagram, we see that the force necessary to move the mass up the incline must be in excess of the component force of gravity trying to pull the mass down the incline. The component of gravity down the incline is always given by $mg\sin\theta$. Resolving the given force into a component parallel to the incline and a component perpendicular to the incline, we find that the force up the incline is, at the same angle, $F\cos\theta$. Thus, in general we would write:

$$F\cos\theta - mg\sin\theta = ma$$

Substituting the known numbers gives $a = 3.17$ m/s^2.

9. **(C)** From the given diagram, we see that the 4-kg mass is accelerating to the right and the 2-kg mass is accelerating up. Thus, we write that the sum of all forces, in each direction, equals ma. In the x-direction, we have (since the tension in the string will try to pull left) $-T + 20 = 4a$. In the y-direction, the tension pulls up against gravity. We therefore write $T - 19.6 = 2a$. Solving for T by eliminating a, we get $T = 19.7$ N.

10. **(A)** We know that the mass accelerates from rest uniformly in 3 s and goes a distance of 3 m. Thus, we can say that

$$d = \frac{1}{2}at^2$$

Substituting the known number gives us an acceleration of 0.67 m/s². On an incline, the normal force is given by $mg\cos\theta$, and so friction can be expressed as $f = \mu mg\cos\theta$. Once again, this force opposes the downward force of gravity parallel to the incline, given by $-mg\sin\theta$. In the downward direction, these two forces are added together and set equal to $-ma$. Thus we write for this case

$$\mu(M)(9.8)\cos 42 - (M)(9.8)\sin 42 = -(M)(0.67)$$

The masses all cancel out, and solving for μ gives 0.8 as the coefficient of kinetic friction.

Free-Response Problems

1. (a) A free-body diagram for each of the two blocks is given below:

Note the two action-reaction pairs here:

- $\vec{N}_1$ is up on the 3-kg mass, and $\vec{N}_1$ is down on the 7-kg mass.
- $\vec{f}_s$ is to the right on the 3-kg mass, and $\vec{f}_s$ is to the left on the 7-kg mass.

(b) The force of static friction between the two blocks is the force responsible for accelerating the 3-kg block to the right (hence the direction of friction in the free-body diagram). On the other hand, kinetic friction opposes the applied force, F, acting on the 7-kg block. Even so, the force F must accelerate both blocks combined. Thus, in the horizontal direction we can write that, for the second law of motion,

$$F - f_k = Ma$$

Also, since $f_k = \mu N$, we can write (since we have horizontal motion, the normal force is equal to the combined weights) $F - (0.35)(10)(9.8) = (10)(5)$. Solving for F, we get $F = 84.3$ N.

(c) Now, for the two blocks together, we know that static friction provides the force needed for the 3-kg block to accelerate at 5 m/s². The normal force ($\vec{N}_1$) on this surface is just the weight of the 3-kg block, as seen in the free-body diagram drawn in part (a). Thus, $f_s = ma$; $\mu N = ma$; $\mu(3)(9.8) = (3)(5)$. From this we get $\mu = 0.51$.

2. (a) A sketch of the situation and the free-body diagram are given below:

Note that a quick tip-to-tail vector addition of our two forces results in a net force towards the center of the circle of travel: $F_N + F_g = F_{net}$

(b) In the coordinates chosen, the horizontal component of the normal force provides the centripetal acceleration. Thus, we can write, in the absence of friction,

$$F_c = ma_c$$

$$N \sin\theta = \frac{mv^2}{r}$$

We can also see from the vertical components that $N\cos\theta = mg$. Thus, eliminating N from both equations gives us

$$\tan\theta = \frac{v^2}{rg}$$

Substituting the known numbers, we find that $\tan\theta = 1.59$ and $\theta = 57.8°$ (rather steep).

(c) Substituting our known value for the angle into the force equation above:

$$N\sin(57.8) = \frac{mv^2}{r}$$

$$N(0.846) = \frac{2,500(25)^2}{40}$$

$$N = 46,000 \text{ N}$$

3. (a) A diagram of the rotor is seen below. Here, the normal force is perpendicular to the person and supplies the inward centripetal force. Friction acts along the wall against gravity, which tends to slide the person down. Thus, the key to stability is to be fast enough to maintain equilibrium.

Free-body
diagram
of person

Substituting in our relationship between velocity and period for circular motion $\left(v = \dfrac{2\pi r}{T} \right)$ into our expression for centripetal forces, we obtain:

$$F_c = ma_c$$

$$F_c = \frac{mv^2}{r}$$

$$N = \frac{4M\pi^2 r}{T^2}$$

In the vertical direction, $f = Mg$; thus

$$f = \mu N = \frac{\mu M 4\pi^2 r}{T^2} = Mg$$

Solving for T, we get

$$T = \sqrt{\frac{\mu 4\pi^2 r}{g}}$$

This makes sense, since if the coefficient of friction is high, the rotation rate can be small, and thus a larger period!

(b) Substituting the known numbers gives $T = 10$ s.

(c) $\omega = \dfrac{2\pi}{T} = \dfrac{6.28}{10} = 0.628$ rad/s.

4. In the frame of reference of the mass, there is an apparent upward force that tends to reduce the apparent weight of the mass at the equator. This effect is very small, and only very sensitive scales can measure it.

5. This is a classic question and is very tricky. The answer is that the action and reaction pairs act on different objects. Thus, the applied force, if it is (or results in) a net force, can cause an object to move in a given frame of reference. The reaction force, however, is experienced by a different object, which may or may not have its own acceleration due to other forces it is experiencing.

Energy

4

KEY CONCEPTS

→ **WORK**

→ **POWER**

→ **KINETIC ENERGY AND THE WORK-ENERGY THEOREM**

→ **POTENTIAL ENERGY AND CONSERVATIVE FORCES**

→ **CONSERVATION OF ENERGY AND SYSTEMS**

WORK

When a force is applied to an object over a displacement, **work** is done on the object. The simplest type of work problem involves a single mass m and a constant force F that causes a displacement d. The angle is between the force and displacement vectors. Work is measured in joules and is a scalar; 1 joule = 1 newton · meter:

$$W = Fd\cos\theta$$

SAMPLE PROBLEM 1

In each scenario below, a force F is applied to a box of mass m that is pushed a distance d horizontally across a frictionless table. If the box is displaced by 20 m, calculate the work W done by the 10 N force indicated.

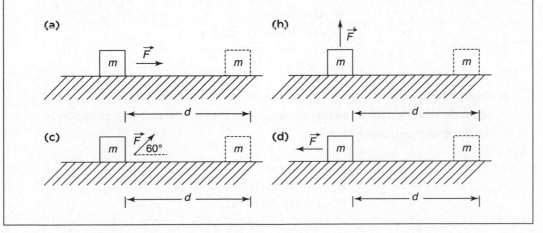

When calculating
work, the mass *m*
of the object
being displaced is
irrelevant.

Solutions

(a) The force and displacement are in the same direction, so all of the force is effective. We use equation $W = Fd$.

$$W = Fd$$
$$W = (10 \text{ N})(20 \text{ m})$$
$$W = 200 \text{ J}$$

Note that $W = Fd$ is a simplified version of $W = (F\cos\theta)d$. The angle is determined by placing both arrows at the origin and measuring the smallest angle between them. If the force and displacement are in the same direction, then $\theta = 0°$. Thus:

$$W = (F\cos\theta)d$$
$$W = (10 \text{ N})(\cos 0°)(20 \text{ m})$$
$$W = (10 \text{ N})(1)(20 \text{ m})$$
$$W = 200 \text{ J}$$

We get the same answer.

(b) The force is not in the same direction as the displacement, so not all of the force is effective. (In fact, none of the force is effective.) We use equation $W = (F\cos\theta)d$. The force is pushing at right angles to the displacement, so $\theta = 90°$:

$$W = (F\cos\theta)d$$
$$W = (10 \text{ N})(\cos 90°)(20 \text{ m})$$
$$W = (10 \text{ N})(0)(20 \text{ m})$$
$$W = 0 \text{ J}$$

(c) Again, the force is not in the same direction as the displacement, so not all of the force is effective. We use equation $W = (F\cos\theta)d$. The force is applied at a 60° angle from the direction of displacement, so $\theta = 60°$:

$$W = (F\cos\theta)d$$
$$W = (10 \text{ N})(\cos 60°)(20 \text{ m})$$
$$W = (10 \text{ N})(0.5)(20 \text{ m})$$
$$W = 100 \text{ J}$$

(d) Again, the force is not in the same direction as the displacement, so we use the equation $W = (F\cos\theta)d$. The force is applied in entirely the opposite direction as the displacement, so $\theta = 180°$:

$$W = (F\cos\theta)d$$
$$W = (10 \text{ N})(\cos 180°)(20 \text{ m})$$
$$W = (10 \text{ N})(-1)(20 \text{ m})$$
$$W = -200 \text{ J}$$

In Sample Problem 1, cases (b) and (d) are the most troublesome. In case (b), no work is done by that particular force. This seems counterintuitive because the box does experience a displacement. Remember that work is not simply "effort" on the part of the force. It is effort that results in a displacement of the object *in the direction of the force*. If the object's displacement is unchanged *in the direction of the force*, no energy was transferred to the object *from that particular force*. The source of the force may be expending energy. Think of yourself holding up a heavy weight at a constant height above the floor. If there is no vertical movement, that energy is not being transferred to the object. Therefore, no work is being done on the object.

In case (d), what does it mean to have negative work? The negative sign tells us that the flow of energy is reversed. Energy is being transferred *from* the mass *to* the source of the force. For example, frictional forces often do negative work, meaning that the mass is losing energy.

SAMPLE PROBLEM 2

Calculate the **total work** (or **net work**) done on object M by the four forces indicated.

$\vec{F_1} = 50$ N
$76°$
$\vec{F_3} = 1$ N ← $\boxed{M}$ → $\vec{F_2} = 7$ N
$\mid \leftarrow d = 2$ m $\rightarrow \mid$
$\vec{F_4} = 20$ N

Solution

Total, or net, work is the sum of the individual works done by each of the individual forces.

$$W_{total} = W_1 + W_2 + W_3 + W_4$$
$$W = (F\cos\theta)d$$
$$W_1 = (50 \text{ N})(\cos 104°)(2 \text{ m})$$
$$W_1 = (10 \text{ N})(-0.24)(2 \text{ m})$$
$$W_1 = -24 \text{ J}$$
$$W_2 = (7 \text{ N})(2 \text{ m})$$
$$W_2 = 14 \text{ J}$$
$$W_3 = (1 \text{ N})(\cos 180°)(2 \text{ m})$$
$$W_3 = (1 \text{ N})(-1)(2)$$
$$W_3 = -2 \text{ J}$$
$$W_4 = (20 \text{ N})(\cos 90°)(2 \text{ m})$$
$$W_4 = (20 \text{ N})(0)(2 \text{ m})$$
$$W_4 = 0$$
$$W_{total} = (-24 \text{ J}) + (14 \text{ J}) + (-2 \text{ J}) + (0 \text{ J})$$
$$W_{total} = -12 \text{ J}$$

REMEMBER

θ is measured relative to the direction of displacement.

In Sample Problem 2, can we compute the net force and then do one work calculation? Try it, and see if you get the same answer. The work done by the net force is equal to the work done by all the individual forces *when you assume that all forces are acting on the center of mass*. Note that this assumption applies in translational motion only but not in rotational motion or deformations of the object.

POWER

When work is done over some amount of time, we can measure the **power**, or the rate at which the work is performed. Power is a scalar with units of watts; 1 watt = 1 joule/sec:

TIP

When a constant force *F* is applied to an object moving at constant velocity *v*, the power *P* is the product of the two values: *P* = *Fv*.

$$P = \frac{W}{t}$$

$$P = \frac{\Delta E}{t}$$

It is an important skill to be able to rewrite any defined unit in terms of the fundamental units of meters, kilograms, seconds, and coulombs. Dimensional analysis of the units of work and power:

$$\text{Work: Joules} = \text{Nm} = (\text{kgm/s}^2)\text{m} = \text{kgm}^2/\text{s}^2$$
$$\text{Power: Watts} = \text{J/s} = \text{Nm/s} = (\text{kgm/s}^2)\text{m/s} = \text{kgm}^2/\text{s}^3$$

Note that the next to last step in the power unit analysis clearly shows units of force times velocity as in the "Tip" on this page.

SAMPLE PROBLEM 3

If the 2-meter displacement in Sample Problem 2 occurred in 4 seconds, how much power was delivered to the object?

Solution

We use the equation $P = \frac{W}{t}$.

$$P = \frac{W}{t}$$

$$P = \frac{-12 \text{ J}}{4 \text{ s}}$$

P = −3 W (The negative sign indicates that the object is losing energy rather than gaining it.)

A 15-kg bucket is raised from the bottom of a well in 2.5 seconds. The well is 8 meters deep, and the bucket is not in motion at the beginning or end of the lift. What was the power exerted while raising the book?

Solution

We use the equation $P = \frac{\Delta E}{t}$, where the energy is gravitational potential energy since this is the only change in the bucket's energy:

$$\Delta E = mg\Delta h = (15)(9.8)(8) = 1176 \text{ J}$$

$$P = \frac{\Delta E}{t} = 1176 \text{ J}/(2.5 \text{ s}) = 470 \text{ J/s} = 470 \text{ W}$$

KINETIC ENERGY AND THE WORK-ENERGY THEOREM

If work is a measure of energy transferred, what does a gain, or loss, of energy mean? One way energy can be easily seen is through an object's motion. This energy of motion is called **kinetic energy**, KE:

$$KE = \frac{1}{2}mv^2$$

A 16-kg object starts at rest and gains 2 J of kinetic energy. Determine the final speed of the object.

Solution

Since the object started at rest, the change in kinetic energy is exactly the kinetic energy of the moving object:

$$KE = \frac{1}{2}mv^2$$

$$2 \text{ J} = \frac{1}{2}(16 \text{ kg})v^2$$

$$\frac{1}{4} = v^2$$

$$0.5 \text{ m/s} = v$$

If we are calculating the net work done by all external forces, the gain or loss of energy W is seen as a change in the speed of the object, which is incorporated in the object's kinetic energy KE. This is the **work-energy theorem**:

$$W_{\text{total}} = \Delta KE$$

A 5-kg object is sliding across a floor at 10 m/s. How much work is done by friction to bring it to a stop?

Solution

The change in kinetic energy is the only calculation we can make:

$$KE_i = \text{initial kinetic energy}$$
$$KE_f = \text{final kinetic energy}$$
$$v_i = \text{initial velocity}$$
$$v_f = \text{final velocity}$$

We are told that $v_i = 10$ m/s. Since the object comes to a stop, we know that $v_f = 0$ m/s:

$$W_{\text{total}} = \Delta KE = KE_f - KE_i$$
$$W_{\text{total}} = \frac{1}{2}mv_f^2 - \frac{1}{2}mv_i^2$$
$$W_{\text{total}} = \frac{1}{2}(16 \text{ kg})(0 \text{ m/s})^2 - \frac{1}{2}(16 \text{ kg})(10 \text{ m/s})^2$$
$$W_{\text{total}} = 0 - 250 \text{ J}$$
$$W_{\text{total}} = -250 \text{ J}$$

The object has lost 250 joules of movement energy as it slowed down. This is the work done by friction: −250 J.

Note that if we knew the distance, we could calculate the force required to bring the object to a stop.

POTENTIAL ENERGY AND CONSERVATIVE FORCES

Potential energy, *PE*, refers to energy stored within a system (between particles that are bound by certain types of forces). The forces that are related to potential energy are called **conservative forces**. There are many forms of potential energy. For the purposes of the AP exam, we are concerned with only three: gravitational, elastic (springs), and electrical. Treat all other forces as nonconservative. Generally speaking, we are concerned only with *changes* in potential energies.

Gravity: The farther two masses are separated, the greater the potential energy stored in the system. The mass that is interacting with Earth close to the planet's surface may change the relationship by changing its height: $PE_{\text{gravity}} = mgh$.

Elastics (e.g., springs): The more a spring is compressed or stretched away from its equilibrium point, the greater the potential energy stored in the system. The stretched or compressed spring has stored potential energy: $PE_{\text{elastic}} = \frac{1}{2}kx^2$. For examples, see Chapter 8.

Electrical: The closer two similar charges are held, or the farther apart two opposite charges are held, the greater the potential energy. Electrical potential energy is given by the equation $\Delta PE_{\text{electrical}} = q\Delta V$. For examples, see Chapter 9.

CONSERVATION OF ENERGY AND SYSTEMS

If only conservative forces are present, the **mechanical energy** (kinetic energy KE plus potential energy PE) is conserved. The **law of conservation of energy** states that the initial mechanical energy must equal the final mechanical energy.

$$\text{Initial Mechanical Energy} = \text{Final Mechanical Energy}$$

$$KE_i + PE_i = KE_f + PE_f$$

SAMPLE PROBLEM 7

A rock falls from a height of 20 meters in a vacuum. How fast is the rock traveling when it reaches the bottom?

Solution

In this problem, we are dealing with the potential energy due to gravity. So $PE = mgh$.

Since there are no nonconservative forces in this problem, we use the law of conservation of energy: $KE_i + PE_i = KE_f + PE_f$.

From the information given, we know that:

$h_i = 20$ m
$v_i = 0$ m/s
$h_f = 0$ m

$$KE_i + PE_i = KE_f + PE_f$$

$$\frac{1}{2}mv_i^2 + mgh_i = \frac{1}{2}mv_f^2 + mgh_f$$

$$\frac{1}{2}(0 \text{ m/s})^2 + (10 \text{ m/s})^2(20 \text{ m}) = \frac{1}{2}v_f^2 + (10 \text{ m/s})^2(0 \text{ m})$$

$$200 \text{ m}^2/\text{s}^2 = \frac{1}{2}v_f^2$$

$$400 \text{ m}^2/\text{s}^2 = v_f^2$$

$$20 \text{ m/s} = v_f$$

Note that the direction of v_f is downward. Also note that the mass m canceled out completely at the start of the problem-solving.

An arrow is shot from the roof of a building 30 meters high at 5 m/s and at an angle of 45 degrees. How fast will the arrow be going when it hits the ground?

Solution

From the information given, we know that:

$h_i = 30$ m
$v_i = 5$ m/s
$h_f = 0$ m

$$KE_i + PE_i = KE_f + PE_f$$

$$\frac{1}{2}mv_i^2 + mgh_i = \frac{1}{2}mv_f^2 + mgh_f$$

$$\frac{1}{2}v_i^2 + gh_i = \frac{1}{2}v_f^2 + gh_f$$

$$\frac{1}{2}v_i^2 + gh_i = \frac{1}{2}v_f^2 + g(0 \text{ m})$$

$$\frac{1}{2}v_i^2 + gh_i = \frac{1}{2}v_f^2$$

$$v_f^2 = v_i^2 + 2gh_i$$

$$v_f^2 = (5 \text{ m/s})^2 + (2)(10 \text{ m/s})^2(30 \text{ m})$$

$$v_f^2 = 25 \text{ m}^2/\text{s}^2 + 600 \text{ m}^2/\text{s}^2$$

$$v_f^2 = 625 \text{ m}^2/\text{s}^2$$

$$v_f = 25 \text{ m/s}$$

A 2-kg mass sliding along a frictionless floor at 3 m/s hits a spring (with $k = 200$ N/m), compresses it, and bounces back. What is the maximum compression of the spring?

Solution

In this problem, we are dealing with the potential energy of a spring. So $PE = \frac{1}{2}kx^2$.

Note that the final state refers to the spring at maximum compression. From the information given, we know:

$m = 2$ kg

$x_i = 0$ m

$v_i = 3$ m/s

$v_f = 0$ m/s

REMEMBER

Velocity is 0 m/s when the spring is at maximum compression ($v = 0$ during any turnaround).

$$KE_i + PE_i = KE_f + PE_f$$

$$\frac{1}{2}mv_i^2 + \frac{1}{2}kx_i^2 = \frac{1}{2}mv_f^2 + \frac{1}{2}kx_f^2$$

$$\frac{1}{2}mv_i^2 + \frac{1}{2}k(0\ m)^2 = \frac{1}{2}m(0\ m/s)^2 + \frac{1}{2}kx_f^2$$

$$\frac{1}{2}mv_i^2 = \frac{1}{2}kx_f^2$$

$$mv_i^2 = kx_f^2$$

$$(2\ kg)(3\ m/s)^2 = (200\ N/m)x_f^2$$

$$x_f^2 = \frac{9}{100}\ m$$

$$x_f - \frac{3}{10} - 0.3\ m$$

If **nonconservative forces** are present (e.g., friction), the work done by those forces must be accounted for when calculating the final kinetic and potential energies. If W_{NC} is the work done by nonconservative forces, the energies must be balanced in the following manner:

$$KE_i + PE_i + W_{NC} = KE_f + PE_f$$

SAMPLE PROBLEM 10

A 2-kg rock falls from a cliff 20 m above the ground. The air friction does −39 J of work during the fall. What is the final speed of the rock?

Solution

Since nonconservative forces are involved, we use equation

$$KE_i + PE_i + W_{NC} = KE_f + PE_f$$

From the information given, we know:

$m = 2\ kg$
$h_i = 20\ m$
$v_i = 0\ m/s$
$h_f = 0\ m$

$$KE_i + PE_i + W_{NC} = KE_f + PE_f$$

$$\frac{1}{2}mv_i^2 + mgh_i + (-39\ J) = \frac{1}{2}mv_f^2 = mgh_f$$

$$\frac{1}{2}(2\ kg)(0\ m/s)^2 + (2\ kg)(10\ m/s^2)(20\ m) - (39\ J) = \frac{1}{2}(2\ kg)v_f^2 + (2\ kg)(10\ m/s^2)(0)$$

$$0\ J + 400\ J - 39\ J = v_f^2 + 0\ J$$

$$361\ J = v_f^2$$

$$19\ m/s = v_f$$

Julia is attempting to calculate the final speed of a dropped ball on Earth using energy consideration. She starts with the work-energy theorem:

$$W = \Delta E$$

She then considers the ball to have dropped over a vertical distance of Δy starting from rest. So Julia first determines the work done by gravity:

$$W = fd = mg\Delta y$$

She then computes the change in energy (the final energy being kinetic and the initial energy being gravitational potential):

$$\Delta E = E_f - E_i = \frac{1}{2}mv^2 - mg\Delta y$$

Julia sets the two equations equal to each other:

$$W = \Delta E$$

$$mg\Delta y = \frac{1}{2}mv^2 - mg\Delta y$$

She solves for v:

$$v = \sqrt{4g\Delta y}$$

From simple kinematics, Julia knows this answer is wrong. Find her error, and explain what she should do to correct it.

Solution

Julia has included gravity twice in her work. First she determined the work done by gravity. Then she included gravitational potential energy in her energy term. She must commit to using either

$$W_{net} = \Delta KE$$

(calculate the work done by gravity but do not use gravitational potential energy)

or

$$W_{NC} = \Delta(KE + PE)$$

(use a zero for the work done by nonconservative forces in this example and then include gravitational potential energy)

Once Julia eliminates an extra usage of gravity, the correct answer in either case will agree with kinematics:

$$v = \sqrt{2g\Delta y}$$

SUMMARY

- Work (measured in joules) is both a force over a distance and a transfer of energy.
- Power (measured in watts) is the rate at which work is done.
- Net work done by all forces to an object is equal to the change in kinetic energy of that object.
- Work done by conservative forces is path independent.
- Potential energy is stored energy within a system due to internal conservative forces. If you take into account only the work done by nonconservative forces to a system, the work-energy theorem becomes:

$$W_{NC} = \Delta KE + \Delta PE$$

- If there are only conservative forces, then the total mechanical energy ($KE + PE$) is conserved.
- If mechanical energy is not conserved in an isolated system, the "missing" energy can be found in the form of internal energies (thermal energy, etc.).

Problem-Solving Strategies for Energy Problems

Given a system, is the external force that is doing work putting joules of energy into kinetic, potential, or internal energy? Generally speaking, the work could be going into any of these. Here are some guidelines:

1. If all forces doing work on the object are treated as external forces (i.e., no potential energies are being used), the net work is going into kinetic energy only ($W_{net} = \Delta KE$).
2. If the only external forces are gravity and/or springs, those forces can be described as potential energies and the mechanical energy is conserved.
3. If the problem has both conservative and nonconservative forces, the problem itself must offer clues as to where the nonconservative forces are putting the energy: $W_{NC} = \Delta KE + \Delta PE$.
4. Often students will assume that mechanical energy is strictly conserved in a problem (i.e., all the joules of energy are to be found in KE and PE). They will overlook a nongravitational, nonspring force (usually friction) in the problem that takes some energy out of KE and PE and puts it into IE (internal energy).

Often a problem can be solved either through kinematics or energy. Usually energy is the quicker, more powerful tool. However, time permitting, you can double-check your answer with simple kinematics as long as the acceleration is constant.

Multiple-Choice

1. Which of the following are the units for the spring constant, k?

    (A) $kg \cdot m^2/s^2$
    (B) $kg \cdot s^2$
    (C) $kg \cdot m/s$
    (D) kg/s^2

2. Which of the following is an expression for mechanical power?

    (A) Ft/m
    (B) $F^2 m/a$
    (C) Fm^2/t
    (D) $F^2 t/m$

3. A pendulum consisting of a mass m attached to a light string of length ℓ is displaced from its rest position, making an angle θ with the vertical. It is then released and allowed to swing freely. Which of the following expressions represents the velocity of the mass when it reaches its lowest position?

    (A) $\sqrt{2g\ell(1-\cos\theta)}$
    (B) $\sqrt{2g\ell(\tan\theta)}$
    (C) $\sqrt{2g\ell(\cos\theta)}$
    (D) $\sqrt{2g\ell(1-\sin\theta)}$

4. An engine maintains constant power on a conveyor belt machine. If the belt's velocity is doubled, the magnitude of its average acceleration

    (A) is doubled
    (B) is quartered
    (C) is halved
    (D) is quadrupled

5. A mass m is moving horizontally along a nearly frictionless floor with velocity v. The mass now encounters a part of the floor that has a coefficient of kinetic friction given by μ. The total distance traveled by the mass before it is slowed by friction to a stop is given by

    (A) $2v^2/\mu g$
    (B) $v^2/2\mu g$
    (C) $2\mu g v^2$
    (D) $\mu v^2/2g$

6. Two unequal masses are dropped simultaneously from the same height. The two masses will experience the same change in

    (A) acceleration
    (B) kinetic energy
    (C) potential energy
    (D) velocity

7. A pendulum that consists of a 2-kg mass swings to a maximum vertical displacement of 17 cm above its rest position. At its lowest point, the kinetic energy of the mass is equal to

(A) 0.33 J
(B) 3.33 J
(C) 33.3 J
(D) 333 J

8. A 0.3-kg mass rests on top of a spring that has been compressed by 0.04 m. Neglect any frictional effects, and consider the spring to be massless. Then, if the spring has a constant k equal to 2,000 N/m, to what height will the mass rise when the system is released?

(A) 1.24 m
(B) 0.75 m
(C) 0.54 m
(D) 1.04 m

9. A box is pulled along a smooth floor by a force F, making an angle θ with the horizontal. As θ increases, the amount of work done to pull the box the same distance, d,

(A) increases
(B) increases and then decreases
(C) remains the same
(D) decreases

10. As the time needed to run up a flight of stairs decreases, the amount of work done against gravity

(A) increases
(B) decreases
(C) remains the same
(D) increases and then decreases

Free-Response

1. A 0.75-kg sphere is dropped through a tall column of liquid. When the sphere has fallen a distance of 2.0 m, it is observed to have a velocity of 5 m/s.

   (a) How much work was done by the frictional "viscosity" of the liquid?
   (b) What is the average force of friction during the displacement of 2.0 m?

2. A 15-kg mass is attached to a massless spring by a light string that passes over a frictionless pulley as shown below. The spring has a force constant of 500 N/m and is unstretched when the mass is released. What is the velocity of the mass when it has fallen a distance of 0.3 m?

$k = 500$ N/m m 15 kg

3. A 1.5-kg block is placed on an incline. The mass is connected to a massless spring by means of a light string passed over a frictionless pulley, as shown below. The spring has a force constant k equal to 100 N/m. The block is released from rest, and the spring is initially unstretched. The block moves down a distance of 16 cm before coming to rest. What is the coefficient of kinetic friction between the block and the surface of the incline?

4. Explain how it might be possible for a moving object to possess, and simultaneously not possess, kinetic energy?

5. When you hold up a 10-kg mass with your arms outstretched, you get tired. However, according to physics, you have not done any work! Explain how this is possible.

6. A car (1200 kg) is traveling up a steep incline (35 degrees with the horizontal) at a speed of 25 MPH. How much power is required?

ANSWERS EXPLAINED

Multiple-Choice Problems

1. **(D)** The units for the spring or force constant are provided by Hooke's law, $F = kx$, and are newtons per meter. Recall that $1 \text{ N} = 1 \text{ kg m/s}^2$, so dividing by m gives kg/s^2.

2. **(D)** Power is equal to work done divided by time. Therefore,

$$P = \frac{W}{t} = \frac{Fd}{t}$$

Using some algebra and kinematics, we see that

$$P = \frac{Fd}{t} = Fv = Fat = F\left(\frac{F}{m}\right)t = \frac{F^2 t}{m}$$

You could also get the answer by verifying which expression has units of joules per second.

3. **(A)** Consider the sketch below of the situation:

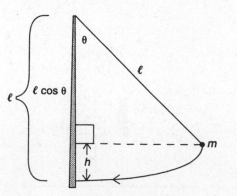

From the geometry of the sketch, note that

$$h = \ell - \ell\cos\theta = \ell(1-\cos\theta)$$

If it is assumed that there is no friction, gravity is the only conservative force acting to do work. Therefore,

$$ME_{bottom} = ME_{top}$$
$$KE_{bottom} = PE_{top}$$
$$\frac{1}{2}mv^2 = mgh$$

and so, at the bottom,

$$v = \sqrt{2gh} = \sqrt{2g\ell(1-\cos\theta)}$$

4. **(C)** Power is equal to the product of the average force applied times the velocity. If the velocity is doubled, and the power is constant, the average force must be halved. Since $F = ma$, the average acceleration of the belt must be halved as well.

5. **(B)** The only applied force is friction, which is doing work to stop the mass. This work is being taken from the initial kinetic energy. For friction, we know that $f = \mu N$, and since the motion is horizontal, $N = mg$. Let x be the distance traveled while stopping; therefore, we can write that $W_f = KE$ and:

$$\frac{1}{2}mv^2 = \mu mgx$$

Solving for x, we get:

$$x = \frac{v^2}{2\mu g}$$

6. **(D)** Objects dropped simultaneously from the same height have the same constant acceleration, which is the change in velocity. The unequal masses will provide for different energies, but the velocity changes will be the same.

7. **(B)** Since we have a conservative system,

$$KE_{bottom} - PE_{top}$$
$$= mgh$$
$$= (2)(9.8)(0.17)$$
$$= 3.33 \text{ J}$$

Remember to change 17 cm to 0.17 m!

8. **(C)** We are dealing with a conservative system, so the initial starting energy is just the potential energy of the compressed spring. This energy supplies the work needed to raise the mass a height h, which is a gain in gravitational potential energy. Thus we equate these two expressions and solve for the height:

$$\frac{1}{2}(2{,}000)(0.4)^2 = (0.3)(9.8)h$$

Thus, $h = 0.54$ m.

9. **(D)** The component of the applied force in the horizontal direction depends on the cosine of the angle. This value decreases with increasing angle. Thus, the work decreases as well.

10. **(C)** The work done is independent of the time or path taken because gravity is a conservative force. The power generated is affected by time, but the work done to run up the stairs remains the same as long as the same mass is raised to the same height.

Free-Response Problems

1. (a) The change in potential energy is a measure of the initial energy and equals $(0.75)(9.8)$ $(2) = 14.7$ J. After the sphere has fallen 2 m, its velocity is 5 m/s, so the kinetic energy is given by

$$KE = \left(\frac{1}{2}\right)(0.75)(5)^2 = 9.375 \text{ J}$$

The work done by friction is due to a nonconservative force that is equal to the difference between the final and initial energies:

$$E_f - E_i = 9.375 \text{ J} - 14.7 \text{ J} = -5.325 \text{ J}$$

Therefore, $W_f = -5.325$ J.

(b) The average frictional force is equal to the work done divided by the displacement of 2 m. Thus

$$f = -2.67 \text{ N}$$

which of course is negative since it opposes the motion.

2. The loss of potential energy is balanced by a gain in elastic potential energy for the spring and in kinetic energy for the falling mass if we assume that the starting energy for the system is zero relative to the starting point for the mass. In the absence of friction, the displacement of the mass is equal to the elongation of the spring. Thus, we can equate our energies and write:

$$0 = -mgh + \frac{1}{2}kx^2 + \frac{1}{2}mv^2$$

If we substitute the known numbers, we get

$$0 = -(15)(9.8)(0.3) + \left(\frac{1}{2}\right)(500)(0.3)^2 + \left(\frac{1}{2}\right)(15)(v)^2$$

Solving for velocity v gives $v = 1.7$ m/s.

3. In this problem, the work done by gravity down the incline is affected by the work done by friction. Together, the net work is applied to stretching the spring by an amount equal to the displacement of the mass. Thus, we can say

$$W_g - W_f = W_s$$

where W_g is the work done by gravity, W_f is the work done by friction, and W_s is the work done to the spring. Hence:

$$mg \sin\theta \, d - \mu \, mg \cos\theta \, d = \frac{1}{2}kx^2$$

Substituting the known numbers, we get:

$$(1.5)(9.8)(0.16)\sin 35 - \mu(1.5)(9.8)(0.16)\cos 35 = \frac{1}{2}(100)(0.16)^2$$

Solving for the coefficient of friction gives $\mu = 0.036$.

4. The motion of an object is always relative to a given frame of reference. If an object is moving relative to one frame, we can envision a second frame, moving with the object, in which it appears to be at rest. Thus, in one frame, the object possesses kinetic energy because of its relative motion. In the second frame, with the object appearing to be at rest, it does not possess kinetic energy.

5. When you hold up a mass, your arm muscles strain against the force of gravity. This requires energy from your body, which makes you feel tired.

6. We will use the power equation when we have a constant force at constant velocity:

$$P = F_v$$

First, however, we must convert the speed of 25 MPH into m/s:

$$25 \text{ miles/hour} \times (1 \text{ hour/3600 seconds}) \times (1609 \text{ meters/1 mile}) = 11 \text{ m/s}$$

Next, we use our knowledge of inclined planes to determine the force the car must be exerting in order to cancel out the down-along-the-incline force of gravity. We know the forces are canceling since the velocity is constant:

$$F_{car} = mg \sin(35) = 1200(9.8)(.574) = 6745 \text{ N}$$

Putting it together:

$$P = F_v = (6745 \text{ N})(11 \text{ m/s}) = 74,000 \text{ W}$$

Gravitation

5

KEY CONCEPTS
→ **NEWTON'S LAW OF UNIVERSAL GRAVITATION**
→ **GRAVITATIONAL ENERGY**

NEWTON'S LAW OF UNIVERSAL GRAVITATION

We already know that all objects falling near the surface of Earth have the same acceleration, given by $g = 9.8$ m/s^2. This is most easily observed by the independence of the period of a pendulum on the mass of the bob. This empirical verification is independent of any "theory" of gravity.

The weight of an object on Earth is given by $F_g = mg$, which represents the magnitude of the force of gravity due to Earth that is acting on the object. From our discussion in Chapter 2, we know that gravity causes a projectile to assume a parabolic path. Isaac Newton, in his book *Principia*, extended the idea of projectile motion to an imaginary situation in which the velocity of the projectile was so great that the object would fall and fall, but the curvature of Earth would bend away and leave the projectile in "orbit." Newton conjectured that this might be the reason why the Moon orbits Earth. This is shown in Figure 5.1.

Figure 5.1

To answer this question, Newton first had to determine the centripetal acceleration of the Moon based on observations from astronomy. He knew the relationship between centripetal acceleration and period. We have written that relationship as

$$a_c = \frac{4\pi^2 r}{T^2}$$

where r is the distance to the Moon (in meters) and T is the orbital period (in seconds). From astronomy we know that $r = 3.8 \times 10^8$ m, and since the Moon orbits Earth in 27.3 days, we have $T = 2.3 \times 10^6$ s. Using these values, we find that the magnitude of $a_c = 2.8 \times 10^{-3}$ m/s^2. Newton realized that some kind of universal gravity was acting as the centripetal force on the Moon.

Using his third law of motion, Newton realized that the force that Earth exerts on the Moon should be exactly equal, but opposite, to the force that the Moon exerts on Earth (observed as tides). This relationship implied that the force should be dependent on the mass of Earth, and this mutual interaction implied that the force of gravity should be proportional to the product of both masses. In other words, Newton's law of gravity could be expressed as

$$F_g = \frac{GM_1 M_2}{r^2}$$

where r is measured from center to center. This is a vector equation in which the force of gravity is directed inward toward the center of mass for the system (in this case, near the center of Earth). Since Earth is many times more massive than the Moon, the Moon orbits Earth, and not vice versa.

The value of G, called the universal gravitational constant, was experimentally determined by Henry Cavendish in 1795. In modern units $G = 6.67 \times 10^{-11}$ N $\cdot$ m^2/kg^2. If we recognize that $F = ma$, and then if we consider M_1 to equal the mass of an object of mass m, and M_2 equal to the mass of Earth, M_E, the acceleration of a mass m near the surface of Earth is given by

$$F_g = ma$$
$$a = \frac{GM_E}{r_E^2}$$

where r_E is the radius of Earth in meters. Using known values for these quantities, we discover that $a = 9.8$ m/s$^2 = g$!

Thus, we have a theory that accounts for the value of the known acceleration due to gravity. In fact, if we replace the mass of Earth by the mass of any other planet, and the radius of Earth by the corresponding radius of the other planet, the above formula allows us to determine the value of g on any planet or astronomical object in the universe! For example, the value of g on the Moon is approximately 1.6 meters per second squared, or about one-sixth the value on Earth. Thus, objects on the Moon weigh one-sixth as much as they do on Earth. This formula also allows us to solve for the acceleration due to gravity at any distance above Earth's surface. Replace r_E with any r greater than r_E.

TIP

Newton's law of universal gravitation is referred to as an inverse square law. For example, if the distance between the masses is doubled, the force between them is one-fourth.

TIP

Notice that the acceleration due to gravity does not depend on the mass of the falling object. This is consistent with Galileo's observations of falling bodies and the period of a pendulum.

Calculate the gravitational force of attraction between a 2,000-kg car and a 12,000-kg truck, separated by 0.5 m.

Solution

We use Newton's law of gravitation:

$$F_g = \frac{GM_1 M_2}{R^2}$$

$$F_g = \frac{(6.67 \times 10^{-11} \text{N} \cdot \text{m}^2/\text{kg}^2)(2,000 \text{ kg})(12,000 \text{ kg})}{(0.5 \text{ m})^2} = 6.4 \times 10^{-3} \text{ N}$$

GRAVITATIONAL ENERGY

In Chapter 4, we saw that the amount of work done (by or against gravity) when vertically displacing a mass is given by the change in the gravitational potential energy:

$$\Delta PE = mg\Delta h$$

This formula is good only for h values that are small when compared to r_E and only at the surface of Earth.

We now know that the value of g is not constant but varies inversely with the square of the distance from the center of Earth. Also, since we want the potential energy to become smaller the closer we get to Earth, we can use the results of the previous section to rewrite the potential energy formula as

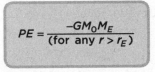

$$PE = \frac{-GM_0 M_E}{(\text{for any } r > r_E)}$$

A graph of universal gravitational potential energy (GPE) as a function of distance is shown in Figure 5.2.

Figure 5.2

As an object gets to higher and higher distances above Earth's surface, the gravitational potential energy does increase like the simplified *mgh*. (It starts out as a large negative number and approaches zero!) The negative values of energy indicate that the object is bound to Earth and can be free only if the object has enough kinetic energy to give it an overall positive mechanical energy.

The "escape velocity" from the gravitational force of Earth can be determined by considering the situation where an object has just reached infinity, with zero final velocity, given some initial velocity at any direction away from the surface of Earth. We designate that escape velocity as v_{esc}, and state that, when the final velocity is zero (at infinity), the total energy must be zero, which implies

$$\frac{1}{2} M_0 v_{esc}^2 = \frac{GM_0 M_E}{r_E}$$

The mass of the object can be eliminated from the relationship, leaving

$$v_{esc} = \sqrt{\frac{2GM_E}{r_E}}$$

THINK ABOUT IT

Notice that both the **escape velocity** and the **orbital velocity** depend on the mass of Earth and not on the mass of the object.

An object can leave the surface of Earth with any velocity. There is, however, one minimum velocity at which, if the spacecraft coasted, it would not fall back to Earth because of gravity.

The orbital velocity can be determined by assuming that we have an approximately circular orbit (see Chapter 2). In this case, we can set the centripetal force equal to the gravitational force:

$$\frac{GM_0 M_E}{r_E^2} = \frac{M_0 v_0^2}{r_E}$$

Eliminating the mass of the object from the equation leaves

$$v_{orbit} = \sqrt{\frac{GM_E}{r_E}}$$

By comparing the two expressions for escape velocity and orbital velocity, we see that

$$v_{esc} = \sqrt{2} v_{orbit}$$

Of course, we could repeat our derivation of orbital speed for any planet of arbitrary mass *M* and any arbitrary orbital radius *r*:

$$v_{orbit} = \sqrt{\frac{GM}{r}}$$

By assuming the orbit is circular, we can substitute in $2\pi r/T$ for v_{orbit}. By setting these two expressions for orbital speed equal to each other, we will be able to derive one of Kepler's original Laws of Planetary Motion, which he codified in 1619. Kepler's third law of planetary motion was that all orbiting objects (orbiting around the same central mass) have the same ratio of T^2/R^3. Although Kepler found this relationship to be the same for all orbiters, he could not explain why, nor did he know what the value represented. But, with the laws of Newton, we can show that the ratio is based on the mass of the central object:

$$2\pi r/T = v_{orbit} = \sqrt{\frac{GM}{r}}$$

Squaring both sides and collecting like terms:

$$\frac{T^2}{r^3} = \frac{4\pi^2}{GM} = \text{Constant}$$

SAMPLE PROBLEM

(a) What is the magnitude of the acceleration due to gravity at an altitude of 400 km above the surface of Earth?

(b) What percentage loss in the weight of an object results?

Solution

(a) The formula for the acceleration due to gravity above Earth's surface is

$$g = \frac{GM_E}{(r_E + h)^2}$$

We have $M_E = 5.98 \times 10^{24}$ kg, $r_E = 6.38 \times 10^6$ m, $h = 400$ km $= 0.4 \times 10^6$ m. Substituting these values, as well as the known value for G given earlier in this chapter, we get $g = 8.67$ m/s^2.

(b) The fractional change in weight is found by comparing the value of g at 400 km to its value at Earth's surface: $8.67/9.8 = 0.885$. Thus, there is an 11.5% loss of weight at that height.

SAMPLE PROBLEM

Three uniform spheres of masses 1 kg, 2 kg, and 4 kg are placed at the corners of a right triangle as shown below. The positions relative to the coordinate system indicated are also shown. What is the magnitude of the resultant gravitational force on the 4-kg mass if we consider that mass to be fixed?

Solution

In this problem, we have to determine separately the force of gravitational attraction between each of the smaller masses and the 4-kg mass. Then we must do a vector sum of these forces.

We begin by calculating the force between the 1-kg mass and the 4-kg mass:

$$F_{1-4} = \frac{(6.67 \times 10^{-11})(1)(4)}{2^2} = 6.67 \times 10^{-11} \text{ N}$$

The direction is to the right and is therefore considered positive.

For the 2-kg mass and the 4-kg mass, the downward direction is taken as negative:

$$F_{2-4} = \frac{(6.67 \times 10^{-11})(2)(4)}{3^2} = 5.93 \times 10^{-11} \text{ N}$$

The magnitude of the resultant force is given by the Pythagorean theorem since F_{1-4} is horizontal and F_{2-4} is vertical.

$$|\vec{F}| = \sqrt{(6.67 \times 10^{-11})^2 + (5.93 \times 10^{-11})^2} = 8.92 \times 10^{-11} \text{ N}$$

SAMPLE PROBLEM

Thomas is studying orbital motion. He is trying to determine what new speed a satellite would have if its altitude above the ground was doubled and if its mass was doubled. He starts by using the equation for orbital velocity:

$$v = \sqrt{\frac{GM}{r}}$$

He then reasons that replacing M with $2M$, and r with $2r$, would leave him with the same expression for orbital speed. He concludes that all satellites orbit with the same speed.

Is Thomas correct? If not, what mistakes has he made? Justify your answer.

Solution

Thomas has made two separate mistakes in his analysis. The first is that the M in the orbital velocity equation is for the mass of the central object. The orbiting object's mass has canceled out in the derivation of this formula and is no longer relevant. So the conclusion that the mass of the satellite does not affect orbital speed is correct (but not for the reason he gave). Thomas's second error is that the r in the equation is from the center of the planet to the center of the satellite. So a doubling of altitude (which is measured from Earth's surface) would be less than a doubling of the actual radius of the original orbit. Generally speaking, satellites at higher altitudes orbit at lower speeds. They do this, however, as a function of orbital radius, not altitude.

SUMMARY

- Newton's law of gravitation states that there is a force of attraction between any two masses and the magnitude of this force is directly proportional to the product of the masses and inversely proportional to the square of the distance between them.
- Newton's law of gravitation is sometimes referred to as an inverse square law similar to Coulomb's law for electrostatics (see Chapter 9).

Problem-Solving Strategies for Gravitation

Keep in mind that gravity is a force and therefore a vector quantity. Distances are measured in meters from the center of masses of each object; therefore, if an object is above the surface of Earth, you must add the height to the radius of Earth. If you remember that the force of gravity is an inverse-square-law relationship, you may be able to deduce an answer from logic rather than algebraic calculations, which take time.

PRACTICE EXERCISES

Multiple-Choice

1. What is the value of g at a height above Earth's surface that is equal to the radius of Earth?

    (A) 9.8 N/kg
    (B) 4.9 N/kg
    (C) 6.93 N/kg
    (D) 2.45 N/kg

2. A planet has half the mass of Earth and half the radius. Compared to the acceleration due to gravity near the surface of Earth, the acceleration of gravity near the surface of this other planet is

    (A) twice as much
    (B) one-fourth as much
    (C) half as much
    (D) the same

3. Which of the following is an expression for the acceleration of gravity at the surface of a planet with uniform density ρ and radius r?

    (A) $G(4\pi\rho/3r^2)$
    (B) $G(4\pi\rho r^2/3)$
    (C) $G(4\pi\rho/3r)$
    (D) $G(4\pi r\rho/3)$

4. What is the orbital velocity of a satellite at a height of 300 km above the surface of Earth? (The mass of Earth is approximately 6×10^{24} kg, and its radius is 6.4×10^6 m.)

(A) 5.42×10^1 m/s

(B) 1.15×10^6 m/s

(C) 7.7×10^3 m/s

(D) 6×10^6 m/s

5. What is the escape velocity from the Moon, given that the mass of the Moon is 7.2×10^{22} kg and its radius is 1.778×10^6 m?

(A) 1.64×10^3 m/s

(B) 2.32×10^3 m/s

(C) 2.69×10^6 m/s

(D) 5.38×10^6 m/s

6. A black hole theoretically has an escape velocity that is greater than or equal to the velocity of light (3×10^8 m/s). If the effective mass of the black hole is equal to the mass of the Sun (2×10^{30} kg), what is the effective "radius" (called the "Schwarzschild radius") of the black hole?

(A) 3×10^3 m

(B) 1.5×10^3 m

(C) 8.9×10^6 m

(D) 4.45×10^6 m

7. What is the gravitational force of attraction between two trucks, each of mass 20,000 kg, separated by a distance of 2 m?

(A) 0.057 N

(B) 0.013 N

(C) 0.0067 N

(D) 1.20 N

8. The gravitational force between two masses is 36 N. If the distance between masses is tripled, the force of gravity will be

(A) the same

(B) 18 N

(C) 9 N

(D) 4 N

Free-Response

1. Find the magnitude of the gravitational field strength g at a point P along the perpendicular bisector between two equal masses, M and M, that are separated by a distance $2b$ as shown below:

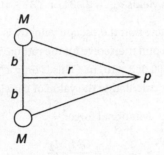

2. Explain why a heavier object near the surface of Earth does not fall faster than a lighter object (Neglect air resistance).

3. Show that the units for g in N/kg are equivalent to m/s². Why is g referred to as the "gravitational field strength"?

4. Explain why objects in orbit appear to be "weightless."

5. Find the orbital radius for a geosynchronous satellite. A geosynchronous satellite is a satellite around Earth that is synchronized, in that its orbital period is 24 hours, just like Earth itself. The mass of Earth is approximately 6.0×10^{24} kg.

ANSWERS EXPLAINED

Multiple-Choice Problems

1. **(D)** The value of g varies inversely with the square of the distance from the center of Earth; therefore, if we double the distance from the center (as in this case), the value of g decreases by one-fourth: $1/4(9.8) = 2.45$. Since $g = F/m$, alternative units are newtons per kilogram.

2. **(A)** If, using the formula for g, we take half the mass, the value decreases by one-half. If we decrease the radius by half, the value will increase by four times. Combining both effects results in an overall increase of two times.

3. **(D)** The formula for g is $g = GM/r^2$. The planet is essentially a sphere of mass M, radius r, and density ρ (with $M = V\rho$; where V is the volume). The volume of the planet is given by $V = 4/3\pi r^3$. Making the substitutions yields:

$$g = G\frac{4\pi r\rho}{3}$$

4. **(C)** The formula for orbital velocity is

$$v_{orbit} = \sqrt{\frac{GM}{r}}$$

where r is the distance from the center of Earth. In this case, we must add 300 km = 300,000 m to the radius of Earth. Thus, $r = 6.7 \times 10^6$ m. Substituting the given values yields $v_{orbit} = 7,728$ or 7.7×10^3 m/s.

5. **(B)** The formula for escape velocity is

$$v_{esc} = \sqrt{\frac{2GM_{Moon}}{r_{Moon}}}$$

Substituting the given values yields $v_{esc} = 2{,}324$ or 2.32×10^3 m/s.

6. **(A)** From question 6, we know that the escape velocity is given by $v_{esc} = \sqrt{2GM/r}$. To find r, we need to square both sides, and then solve for the radius. This yields $r = 2GM/v^2$. Substituting the given values yields $r = 3{,}000$ or 3×10^3 m. (This is only a "theoretical" size for the black hole. As an interesting exercise, try calculating the value of g on such an object!)

7. **(C)** We use the formula for gravitational force:

$$F = \frac{GM_1M_2}{r^2}$$

Substituting the given values (don't forget to square the distance!) yields $F = 0.0067$ N.

8. **(D)** The force of gravity is an inverse-square-law relationship. This means that, as the distance is tripled, the force is decreased by one-ninth. One-ninth of 36 N is 4 N.

Free-Response Problems

1. Let's designate as g_1 the field strength at P caused by the top mass and designate as g_2 the field strength at point P due to the bottom mass. Both of these field strengths are accelerations and therefore vectors. The distance from point P to the line connecting the masses is r, and the midpoint distance connecting the masses is b. Therefore, the distance from point P to each mass is given by the Pythagorean theorem and is equal to $\sqrt{r^2+b^2}$.

The direction of each acceleration g is directed toward each mass from point P. Since each mass is identical and the distance to each mass is the same, the angles formed by the vectors to the x-axis are the same. Let's call each angle θ such that $\tan\theta = b/r$ in magnitude.

From the above analysis we can conclude that

$$g_1 = g_2 = \frac{GM}{r^2+b^2}$$

The vector components of each field strength result in a symmetrical cancellation of the y-components. This is true because the direction of g_1 is toward the upper left, and thus its x-component is directed to the left and its y-component is directed upward. Field strength g_2 has an x-component that is also directed to the left and equal in magnitude to the x-component of g_1. The y-component of g_2 is directed downward and is also equal in magnitude to the y-component of g_1. Since these two vectors are equal and opposite, they will sum to zero and will not contribute to the net resultant field (which is just directed horizontally to the left).

What remains to be done is to determine the expression for the x-component of g_1 or g_2 and then multiply by 2. Since $g_{1x} = g_1\cos\theta$, we can see from the geometry that

$$\cos\theta = \frac{r}{\sqrt{r^2+b^2}}$$

Combining results gives

$$g_{\text{net}} = \frac{2GMr}{(r^2 + b^2)^{3/2}}$$

2. The acceleration due to gravity near the surface of Earth is given by

$$g = \frac{GM_E}{r^2}$$

Hence, this acceleration is independent of the mass of an object.

3. We know that $F = ma$, and for gravity, $F = mg$. Hence,

$$g = \frac{F}{m} = \text{N/kg}$$

But we also know that $N = \text{kg} \cdot \text{m/s}^2$. Thus, in units,

$$\text{N/kg} = \text{m/s}^2$$

A gravitational field measures the amount of force per unit mass. Thus, g is referred to as the gravitational field strength since $g = F/m$.

4. Objects in orbit are in free fall. Thus, all objects fall together and appear to be weightless. A person in free fall therefore does not experience any contact forces and does not feel his or her own weight.

5. We know the orbital period of this satellite must be 24 hours:

$$T = 24 \text{ hrs} \times (3600 \text{ sec/hour}) = 86{,}400 \text{ s}$$

Using Newton's version of Kepler's third law:

$$\frac{T^2}{r^3} = \frac{4\pi^2}{GM} = \text{Constant}$$

Knowing T for our satellite, the mass of the Earth (6.0×10^{24} kg), and the universal gravitational constant (G), we can solve for r:

$$r = (GMT^2/4\pi^2)^{1/3}$$
$$r = 4.2 \times 10^7 \text{ m}$$

Impacts and Linear Momentum

6

INTERNAL AND EXTERNAL FORCES

Consider a system of two blocks with masses m and M ($M > m$). If the blocks were to collide, the forces of impact would be equal and opposite. However, because of the different masses, the response to these forces (i.e., the changes in velocity) would not be equal. In the absence of any outside or external forces acting on the objects (such as friction or gravity), we say that the impact forces are internal.

To Newton, the "quantity of motion" discussed in his book *Principia* was the product of an object's mass and velocity. This quantity, called **linear momentum** or just **momentum**, is a vector quantity having units of kilogram · meters per second (kg · m/s). Algebraically, momentum is designated by the letter $\vec{p}$, such that $\vec{p} = m\vec{v}$.

To understand Newton's rationale, consider the action of trying to change the motion of a moving object. Do not confuse this with the inertia of the object; here, both the mass and the velocity are important. Consider, for example, that a truck moving at a slow 1 meter per second can still inflict a large amount of damage because of its mass. Also, a small bullet, having a mass of perhaps 1 gram or less, does incredible damage because of its high velocity. In each case (see Figure 6.1), the damage is the result of a force of impact when the object is intercepted by something else. Let's now consider the nature of impact forces.

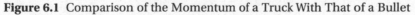

Figure 6.1 Comparison of the Momentum of a Truck With That of a Bullet

IMPACT FORCES AND MOMENTUM CHANGES

Consider a mass m moving with a velocity v in some frame of reference. If the mass is subjected to some external forces, then, by Newton's second law of motion, we can write that $\sum F = ma$. The vector sum of all the forces, referred to as the **net force**, is responsible for changing the velocity of the motion (in magnitude and/or direction).

If we recall the definition of acceleration as the rate of change or velocity, we can rewrite the second law of motion as

$$\vec{F}_{net} = m\left(\frac{\Delta \vec{v}}{\Delta \vec{t}}\right)$$

This expression is also a vector equation and is equivalent to the second law of motion. If we make the assumption that the mass of the object is not changing, we can again rewrite the second law in the form

$$\vec{F}_{net} = \frac{\Delta m \vec{v}}{\Delta t} = \frac{\Delta \vec{p}}{\Delta \vec{t}}$$

This expression means that the net external force acting on an object is equal to the rate of change of momentum of the object and is another alternative form of Newton's second law of motion. This **change in momentum** is a vector quantity in the same direction as the net force applied. Since the time interval is just a scalar quantity, we can multiply both sides by Δt to get

$$\vec{F}_{net}\,\Delta t = \Delta \vec{p} = m\Delta \vec{v} = m\vec{v}_f - m\vec{v}_i$$

The quantity, $\vec{F}_{net}\Delta t$, called the **impulse**, represents the effect of a force acting on a mass during a time interval Δt, and is likewise a vector quantity. From this expression, it can be stated that the impulse applied to an object is equal to the change in momentum of the object.

Another way to consider impulse is to look at a graph of force versus time for a continuously varying force (see Figure 6.2).

The area under this curve is a measure of the impulse in units of newton $\cdot$ seconds (N $\cdot$ s). Another way to view this concept is to identify the average force, $\vec{F}_{avg}$, such that the area of the rectangle formed by the average force is equal in area to the entire curve. This is more manageable algebraically, and we can write

$$\vec{F}_{avg}\Delta t = \Delta \vec{p}$$

A 2,000-kg car is traveling at 20 m/s and stops moving over a 10-s period. What was the magnitude of the average braking force?

Solution

We know that

$$\vec{F} = \frac{m\Delta\vec{v}}{\Delta t}$$

Substituting, we obtain

$$F = \frac{(2,000 \text{ kg})(20 \text{ m/s})}{10 \text{ s}} = 4,000 \text{ N}$$

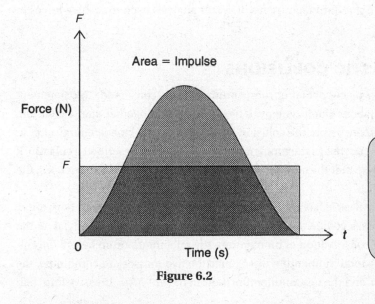

Area = Impulse

Force (N)

F

0

Time (s)

Figure 6.2

How do airbags work? After all, the same impulse must be supplied to the car's occupants during the crash. The airbag lengthens the time interval over which the force is applied. The area under the curve is the same. However, it is wider and not as high (lower force).

THE LAW OF CONSERVATION OF LINEAR MOMENTUM

Newton's third law of motion states that for every action there is an equal but opposite reaction. This reaction force is present whenever we have an interaction between two objects in the universe. Suppose we have two masses, m_1 and m_2, that are approaching each other along a horizontal frictionless surface. Let $\vec{F}_{12}$ be the force that m_1 exerts on m_2, and let $\vec{F}_{21}$ be the force that m_2 exerts on m_1. According to Newton's law, these forces must be equal and opposite; that is, $\vec{F}_{12} = -\vec{F}_{21}$.

Rewriting this expression as $\vec{F}_{12} + \vec{F}_{21} = 0$ leads to an interesting implication. Since each force is a measure of the rate of change of momentum for that object, we can write

$$\vec{F}_{12} = \frac{\Delta\vec{p}_1}{\Delta t} \quad \text{and} \quad \vec{F}_{21} = \frac{\Delta\vec{p}_2}{\Delta t}$$

Therefore:

$$\frac{\Delta\vec{p}_1}{\Delta t} + \frac{\Delta\vec{p}_2}{\Delta t} = 0 \quad \text{and} \quad \frac{\Delta(\vec{p}_1 + \vec{p}_2)}{\Delta t} = 0$$

The change in the sum of the momenta is therefore zero, implying that the sum of the total momentum for the system $(\vec{p}_1 + \vec{p}_2)$ is a constant all the time. This conclusion is called the **law of conservation of linear momentum**, and we say simply that the momentum is conserved.

Here is another way of writing this conservation statement in a general form for any two masses (after separating all initial and final terms):

$$m_1\vec{v}_{1i} + m_2\vec{v}_{2i} = m_1\vec{v}_{1f} + m_2\vec{v}_{2f}$$

Extension of the law of conservation of momentum to two or three dimensions involves the recognition that momentum is a vector quantity. Given two masses moving in a plane relative to a coordinate system, conservation of momentum must hold simultaneously in both the horizontal and vertical directions. These vector components of momentum can be calculated using the standard techniques of vector analysis used to resolve forces in components.

ELASTIC AND INELASTIC COLLISIONS

During any collision between two pieces of matter, momentum is always conserved. This statement is not, however, necessarily true about kinetic energy. If two masses stick together after a collision, it is observed that the kinetic energy before the collision is not equal to the kinetic energy after it. If the kinetic energy is conserved as well as the momentum, the collision is described as **elastic**. If the kinetic energy is not conserved after the collision (e.g., energy being lost to heat or friction), the collision is described as **inelastic**.

TIP

In an elastic collision, the kinetic energy of the system is conserved.

As an example, suppose that a mass m has a velocity $\vec{v}$, while mass M is at rest along a horizontal, frictionless surface. The two masses collide and stick together. What is the final velocity, $\vec{u}$, of the system? According to the law of conservation of momentum, the total momentum before the collision $(m\vec{v})$ must be equal to the total momentum after. Since the two masses are combining, the new mass of the system is $M + m$, and the new momentum is given by $(m + M)\vec{u}$. Thus, we find that:

$$\vec{u} = \frac{m\vec{v}}{m+M}$$

The initial kinetic energy is $(1/2)mv^2$. The final kinetic energy is given by

$$KE_f = \frac{1}{2}(m+M)\vec{u}^2 = \frac{1}{2}\left(\frac{m^2\vec{v}^2}{m+M}\right)$$

which is, of course, not equal to the initial kinetic energy.

TIP

On the AP Physics 1 exam, you may be asked to set up, but not solve, a system of equations. Your understanding of physics is thus demonstrated without having to spend the time doing the algebra!

If we have an elastic collision, we write both conservation laws to get two equations involving the velocities of the masses before and after the collision:

$$m_1\vec{v}_{1i} + m_2\vec{v}_{2i} = m_1\vec{v}_{1f} + m_2\vec{v}_{2f}$$
$$\frac{1}{2}m_1v_{1i}^2 + \frac{1}{2}m_2v_{2i}^2 = \frac{1}{2}m_1v_{1f}^2 + \frac{1}{2}m_2v_{2f}^2$$

If we cancel out the factor $(1/2)$ in the second equation, and collect the expressions for each mass on each side, we can rewrite the two equations as

$$m_1(\vec{v}_{1i} - \vec{v}_{1f}) = m_2(\vec{v}_{2f} - \vec{v}_{2i})$$

$$m_1(\vec{v}_{1i}^2 - \vec{v}_{1f}^2) = m_2(\vec{v}_{2f}^2 - \vec{v}_{2i}^2)$$

The second equation is factorable and so can be simplified. Again rewriting the expressions, we get

$$m_1(\vec{v}_{1i} - \vec{v}_{1f}) = m_2(\vec{v}_{2f} - \vec{v}_{2i})$$

$$m_1(\vec{v}_{1i} + \vec{v}_{1f})(\vec{v}_{1i} - \vec{v}_{1f}) = m_2(\vec{v}_{2f} + \vec{v}_{2i})(\vec{v}_{2f} - \vec{v}_{2i})$$

If we take the ratio of the two expressions, we arrive at an interesting result after collecting terms:

$$\vec{v}_{1i} - \vec{v}_{2i} = -(\vec{v}_{1f} - \vec{v}_{2f})$$

This expression states that the relative velocity between the masses before the elastic collision is equal and opposite to the relative velocity between the masses after the elastic collision! Thus, the characteristic "rebounding" can be observed during elastic collisions.

SAMPLE PROBLEM

A mass of 2 kg is moving at a speed of 10 m/s along a horizontal, frictionless surface. It collides, and sticks, with a 3-kg mass moving in the same direction at 5 m/s.

(a) What is the final velocity of the system after the collision?

(b) What percentage of kinetic energy was lost in the collision?

Solution

(a) Using conservation of momentum, we see that

$$m_1\vec{v}_1 + m_2\vec{v}_2 = (m_1 + m_2)\vec{v}_f$$

$$(2 \text{ kg})(10 \text{ m/s}) + (3 \text{ kg})(5 \text{ m/s}) = (5 \text{ kg})\vec{v}_f$$

$$(35 \text{ kg} \cdot \text{m/s}) = (5 \text{ kg})\vec{v}_f$$

$$\vec{v}_f = 7 \text{ m/s}$$

(b) The initial kinetic energy is given by

$$\frac{1}{2}m_1v_1^2 + \frac{1}{2}m_2v_2^2 = KE_i$$

$$\frac{1}{2}(2 \text{ kg})(10 \text{ m/s})^2 + \frac{1}{2}(3 \text{ kg})(5 \text{ m/s})^2 = 137.5 \text{ J}$$

The final kinetic energy is given by

$$\frac{1}{2}(m_1 + m_2)v_f^2 = KE_f$$

$$\frac{1}{2}(5 \text{ kg})(7 \text{ m/s})^2 = 122.5 \text{ J}$$

We now take the ratio of the final kinetic energy and the initial kinetic energy:

$$(122.5 \text{ J})/(137.5 \text{ J}) = 0.89 \rightarrow 89\% \text{ left over } KE$$

Therefore, 11% of the initial kinetic energy was lost in the collision.

Assuming the two objects collide completely inelastically (they stick together), rank the magnitude of the momenta of the product of these collisions.

(A) 2v → 3m   v → 2m

(B) 3v → 5m   3v ← 2m

(C) 2v ← 2m   6v ← m

(D) 5v → 3m   3v ← 5m

Follow-up: Which collision will release the most energy?

Solution

Simply sum the momenta of object A and object B before the collision. (Keep track of negative signs!) Since momentum is conserved, the highest magnitude (absolute value) total will be the highest after the collision as well.

(A) $6mv + 2mv = 8mv$

(B) $15mv - 6mv = 9mv$

(C) $-4mv - 6mv = -10mv$

(D) $15mv - 15mv = 0$

So, from highest magnitude of momentum to lowest:

$$C \rightarrow B \rightarrow A \rightarrow D$$

Follow-up answer: Crash D has the greatest loss of kinetic energy since all of the kinetic energy is lost during its collision. So crash D will release the greatest amount of energy.

TIP

Note that the AP Physics 1 exam is unlikely to have a calculation question that would require direct use of this formula, but you should understand the concepts.

CENTER OF MASS

The center of mass of a system is a very useful concept. Conceptually, the center of mass is the point in a system about which all the mass is balanced. It is found by taking the weighted average of all points of mass along each axis. For example, along the x-axis:

$$x_{cm} = \sum m_i x_i / \sum m_i$$

It is the center of mass of an object to which we apply our forces in our free-body diagrams. The center of mass of an object is actually located at the positions specified in all our physics problems. For example, a real-world, unsymmetrical object may not appear to travel in a parabola when undergoing projectile motion as it twists and turns during its flight. However, its center of mass is following the parabolic path.

Figure 6.3

When thinking of the momentum of a system of interacting particles, the overall momentum of the system obeys all of Newton's laws relative to external forces. For example, a mass of gas consisting of many moving and interacting molecules can be modeled as being attracted gravitationally to Earth by placing the total mass of the gas cloud at its center of mass.

SUMMARY

- Momentum is a vector quantity equal to $m\bar{v}$.
- The impulse ($\bar{F}\Delta t$) is found by taking the area under a graph of force versus time.
- Impulse is equal to the change in momentum ($\bar{F}\Delta t = m\Delta\bar{v}$).
- Force is equal to the rate of change of momentum.
- In an isolated system, the total momentum is conserved.
- In an elastic collision, the kinetic energy of the system is conserved.
- In an inelastic collision (typically, but not necessarily) where masses stick together, kinetic energy is lost (this energy loss typically transforms into heat).

Problem-Solving Strategies for Impacts and Linear Momentum

In any isolated system, momentum is always conserved. Therefore, when you read a problem that does not state explicitly that momentum is involved, you can safely assume, in the case of collision or an impact, that total momentum is conserved, and you should write the equations for the conservation of momentum. The kinetic energy, however, is not necessarily conserved unless the collision is elastic. In summary, you should:

1. Decide whether an impact or a collision is involved. If there is a collision, observe whether it is elastic or inelastic.
2. Remember that, if the collision is inelastic, masses will usually stick together, so be sure to determine the new combined mass.
3. If the collision is elastic, write the equations for the conservation of both momentum and kinetic energy.
4. Remember that the impulse given to a mass is equal to its change in momentum. The change in momentum is a vector quantity in the same direction as the impulse or net force.
5. Be sure to take into account algebraically any reversal of directions, and remember the sign conventions for left, right, up, and down motions.
6. Keep in mind that motions may be simpler if studied in the center-of-mass frame of reference. In such a frame, one considers the motion of the center of mass, as well as the motions of mass particles relative to the center of mass.
7. Remember that in two dimensions the center of mass will follow a smooth path after an internal explosion since the forces involved were internal and the initial momentum in that frame was zero. For example, if a projectile launched at an angle experiences an explosion mid-flight, the center of mass of all the debris still follows the original parabolic trajectory.

Multiple-Choice

1. Which of the following expressions, where p represents the linear momentum of the particle, is equivalent to the kinetic energy of a moving particle?

   (A) mp^2
   (B) $m^2/2p$
   (C) $2p/m$
   (D) $p^2/2m$

2. Two carts having masses 1.5 kg and 0.7 kg, respectively, are initially at rest and are held together by a compressed massless spring. When released, the 1.5-kg cart moves to the left with a velocity of 7 m/s. What is the velocity and direction of the 0.7-kg cart?

   (A) 15 m/s right
   (B) 15 m/s left
   (C) 7 m/s left
   (D) 7 m/s right

3. The product of an object's instantaneous momentum and its acceleration is equal to its

   (A) applied force
   (B) kinetic energy
   (C) power output
   (D) net force

4. A ball with a mass of 0.15 kg has a velocity of 5 m/s. It strikes a wall perpendicularly and bounces off straight back with a velocity of 3 m/s. The ball underwent a change in momentum equal to

   (A) $0.30 \text{ kg} \cdot \text{m/s}$
   (B) $1.20 \text{ kg} \cdot \text{m/s}$
   (C) $0.15 \text{ kg} \cdot \text{m/s}$
   (D) $5 \text{ kg} \cdot \text{m/s}$

5. What braking force is supplied to a 3,000-kg car traveling with a velocity of 35 m/s that is stopped in 12 s?

   (A) 29,400 N
   (B) 3,000 N
   (C) 8,750 N
   (D) 105,000 N

6. A 0.1-kg baseball is thrown with a velocity of 35 m/s. The batter hits it straight back with a velocity of 60 m/s. What is the magnitude of the average impulse exerted on the ball by the bat?

(A) 3.5 N · s

(B) 2.5 N · s

(C) 7.5 N · s

(D) 9.5 N · s

7. A 1-kg object is moving with a velocity of 6 m/s to the right. It collides and sticks to a 2-kg object moving with a velocity of 3 m/s in the same direction. How much kinetic energy was lost in the collision?

(A) 1.5 J

(B) 2 J

(C) 2.5 J

(D) 3 J

8. A 2-kg mass moving with a velocity of 7 m/s collides elastically with a 4-kg mass moving in the opposite direction at 4 m/s. The 2-kg mass reverses direction after the collision and has a new velocity of 3 m/s. What is the new velocity of the 4-kg mass?

(A) −1 m/s

(B) 1 m/s

(C) 6 m/s

(D) 4 m/s

9. A mass m is attached to a massless spring with a force constant k. The mass rests on a horizontal, frictionless surface. The system is compressed a distance x from the spring's initial position and then released. The momentum of the mass when the spring passes its equilibrium position is given by

(A) $x\sqrt{mk}$

(B) $x\sqrt{k/m}$

(C) $x\sqrt{m/k}$

(D) $x\sqrt{k^2m}$

10. During an inelastic collision between two balls, which of the following statements is correct?

(A) Both momentum and kinetic energy are conserved.

(B) Momentum is conserved, but kinetic energy is not conserved.

(C) Momentum is not conserved, but kinetic energy is conserved.

(D) Neither momentum nor kinetic energy is conserved.

Free-Response

1. Two blocks with masses 1 kg and 4 kg, respectively, are moving on a horizontal, frictionless surface. The 1-kg block has a velocity of 12 m/s, and the 4-kg block is ahead of it, moving at 4 m/s, as shown in the diagram below. The 4-kg block has a massless spring attached to the end facing the 1-kg block. The spring has a force constant k equal to 1,000 N/m.

   (a) What is the maximum compression of the spring after the collision?
   (b) What are the final velocities of the blocks after the collision has taken place?

2. A 0.4-kg disk is initially at rest on a frictionless, horizontal surface. It is hit by a 0.1-kg disk moving horizontally with a velocity of 4 m/s. After the collision, the 0.1-kg disk has a velocity of 2 m/s at an angle of $43°$ to the positive x-axis.

   (a) Determine the velocity and direction of the 0.4-kg disk after the collision.
   (b) Determine the amount of kinetic energy lost in the collision.

3. A 50-kg girl stands on a platform with wheels on a frictionless, horizontal surface as shown below. The platform has a total mass of 1,000 kg and is attached to a massless spring with a force constant k equal to 1,000 N/m. The girl throws a 1-kg ball with an initial velocity of 35 m/s at an angle of $30°$ to the horizontal.

   (a) What is the recoil velocity of the platform-girl system?
   (b) What is the elongation of the spring?

4. (a) Can an object have energy without having momentum? Explain.
   (b) Can an object have momentum without having energy? Explain.

5. Explain why there is more danger when you fall and bounce as opposed to falling without bouncing.

6. A cart of mass M is moving with a constant velocity v to the right. A mass m is dropped vertically onto it, and it is observed that the new velocity is less than the original velocity. Explain what has happened in terms of energy, forces, and conservation of momentum (as viewed from different frames of reference).

ANSWERS EXPLAINED

Multiple-Choice Problems

1. **(D)** If we multiply the formula for kinetic energy by the ratio m/m, we see that the formula for kinetic energy becomes

$$KE = \left(\frac{1}{2m}\right)(m^2)(v^2) = \frac{p^2}{2m}$$

2. **(A)** Momentum is conserved, so $(1.5)(7) = 0.7v$. Thus, $v = 15$ m/s. The direction is to the right since in a recoil the masses go in opposite directions.

3. **(C)** An object's instantaneous momentum times its acceleration will equal its power output in units of joules per second or watts.

4. **(B)** The change in momentum is a vector quantity. The rebound velocity is in the opposite direction, so $\Delta v = 5 - (-3) = 8$ m/s. The change in momentum is

$$\Delta p = (0.15)(8) = 1.20 \text{ kg} \cdot \text{m/s}$$

5. **(C)** The formula is $F\Delta t = m\Delta v$. Solving for the force, we get

$$F = \frac{(3,000)(35)}{12} = 8,750 \text{ N}$$

6. **(D)** Impulse is equal to the change in momentum, which is

$$\Delta p = |0.1[-60 - 35]| = 9.5 \text{ N} \cdot \text{s}$$

because of the change in direction of the ball.

7. **(D)** First, we find the final velocity of this inelastic collision. Momentum is conserved, so we can write $(1)(6) + (2)(3) = (3)v'$ since both objects are moving in the same direction. Thus, $v' = 4$ m/s. The initial kinetic energy of the 1-kg object is 18 J, while the initial kinetic energy of the 2-kg mass is 9 J. Thus, the total initial kinetic energy is 27 J. After the collision, the combined 3-kg object has a velocity of 4 m/s and a final kinetic energy of 24 J. Thus, 3 J of kinetic energy has been lost.

8. **(B)** Momentum is conserved in this elastic collision but the directions are opposite, so we must be careful with negative signs. We therefore write $(2)(7) - (4)(4) = -(2)(3) + 4v'$ and get $v' = 1$ m/s.

9. **(A)** We set the two energy equations equal to solve for the velocity at the equilibrium position. Thus

$$\frac{1}{2}kx^2 = \frac{1}{2}mv^2$$

since no gravitational potential energy is involved, and we write that the velocity is $v = x\sqrt{k/m}$. Now momentum $p = mv$, so we multiply by m and factor the "mass" back under the radical sign, where it is squared so that we get $p = x\sqrt{mk}$.

10. **(B)** Momentum is always conserved in an inelastic collision, and kinetic energy is not conserved because the objects stick together.

Free-Response Problems

1. (a) Upon impact, the spring becomes compressed, but both blocks are still in motion. Therefore, for an instant, we have an inelastic collision, and momentum is, of course, conserved. Thus, we can write for the moment of impact

$$m_1 \vec{v}_{1i} + m_2 \vec{v}_{2i} = (m_1 + m_2) \vec{v}_f$$

Solving for the final velocity, we get $\vec{v}_f = 53.6$ m/s. Using the initial values for the velocities, we find that the initial kinetic energies are 72 J and 32 J for the 1-kg and 4-kg blocks, respectively. Thus, the total initial kinetic energy is 104 J. Using the final velocity of 5.6 m/s and the combined mass of 5 kg, we get a final kinetic energy of 78.4 J. The difference in kinetic energy of 25.6 J is used to compress the spring in this inelastic collision. Using the formula for the work done against a spring, $(1/2)kx^2$, we get $x = 0.05$ m for the maximum compression of the spring after the collision.

(b) After the collision has taken place, the two blocks again separate. If we treat the situation as elastic (since we assume that the work done to compress the spring will be used by the spring in rebounding), we can use the initial velocities as a "before" condition for momentum and kinetic energy. We then seek to solve for the two final velocities of the blocks after the rebound has taken place. To find both velocities, we need two equations, and we use the conservation of kinetic energy in this case to assist us. Thus, we write

$$m_1 \vec{v}_{1i} + m_2 \vec{v}_{2i} = m_1 \vec{v}_{1f} + m_2 \vec{v}_{2f}$$

and for the kinetic energy

$$\frac{1}{2} m_1 v_{1i}^2 + \frac{1}{2} m_2 v_{2i}^2 = \frac{1}{2} m_1 v_{1f}^2 + \frac{1}{2} m_2 v_{2f}^2$$

Substituting the known numbers, we obtain for the momentum $28 = v_{1f} + 4v_{2f}$ and for the kinetic energy we get $208 = v_{1f}^2 + 4v_{2f}^2$.

If we solve for v_{1f} from the momentum equation, square the result, and substitute into the kinetic energy equation, we obtain a factorable quadratic equation for v_{2f}:

$$v_{2f}^2 - 11.2 v_{2f} + 28.2 = (v_{2f} - 7.2)(v_{2f} - 4) = 0$$

Of the two choices for the second final velocity, only one provides a physically meaningful set of solutions. This is so because the first mass must rebound, and hence its final velocity must be negative. Our final answers are therefore $v_{2f} = 7.2$ m/s and $v_{1f} = -0.8$ m/s.

2. (a) This is a two-dimensional collision, and we consider the conservation of momentum in each direction, before and after. In the x-direction, we have only the initial momentum of the 0.1-kg disk with a velocity of 4 m/s. After the collision, the 0.1-kg disk has an x-component of momentum given by $(0.1)(2)\cos 43$.

The 0.4-kg disk had initially zero momentum and now has some unknown velocity at some unknown angle, θ, to the x-axis. Let us assume that the angle is below the x-axis, so that the x-component will be positive and the y-component negative. If our assumption is correct, we will get a positive answer for θ. A negative answer will let us know that the assumption is incorrect. The x-component of final momentum for the 0.4-kg disk is given by $(0.4)v_{2f}\cos\theta$, and the y-component by $-(0.4)v_{2f}\sin\theta$. We can therefore write for the x-direction:

$$(0.1)(4)+0=(0.1)(2)\cos 43+(0.4)v_{2f}\cos\theta$$

and for the y-direction:

$$0=(0.1)(2)\sin 43-(0.4)v_{2f}\sin\theta$$

Solving for the angles and velocities in each, we get the following two equations:

$$v_{2f}\sin\theta = 0.34 \quad \text{and} \quad v_{2f}\cos\theta = 0.635$$

Taking the ratio gives $\tan\theta = 0.535$ and $\theta = 28°$. Substituting this angle gives us the final velocity for the 0.4-kg disk: 0.72 m/s.

(b) To find the loss of kinetic energy, we calculate the total initial and final kinetic energies. Using the given data, we find that the initial kinetic energy is 0.8 J. Using our final data, we get a total final kinetic energy of 0.3 J. Thus, 0.5 J of kinetic energy has been lost.

3. (a) In this problem, the x-component of the velocity provides the impulse to elongate the spring using recoil. Thus, we have that the velocity of the ball is $35\cos 30 = 30.31$ m/s. Using conservation of momentum, we state that $(1)(30.31)=(1,050)v'$. Thus, $v'=0.0289$ m/s in the negative x-direction, so the recoil velocity of the platform-girl system can be expressed as $v' = -0.0289$ m/s.

$$\left(\frac{1}{2}\right)(1,050)(0.0289)^2 =\left(\frac{1}{2}\right)(1,000)x^2$$

(b) Solving for x gives $x = 0.0296$ m for the elongation of the spring.

4. (a) In a given frame of reference, an object can appear to be at rest and possess potential energy. In this frame, the object has zero velocity and hence its momentum is zero.

(b) If an object has nonzero momentum, it must have nonzero velocity, which means it must have kinetic energy.

5. When an object bounces, an additional upward force (impulse) is given to it. This extra force, which provides for an additional change in momentum, can be dangerous if it is large enough.

6. First, if the mass is placed vertically onto the object, the increase in mass appears to lower the velocity since no force was acting in the direction of motion. We can also state that since the object did not have any horizontal motion, the friction between the new mass and the object must act to accelerate the mass m, and this energy comes from the moving object. From the mass M frame of reference, it appears as though the mass m is coming toward it, providing a backward force which will slow down the object.

Rotational Motion

7

KEY CONCEPTS

→ **PARALLEL FORCES AND MOMENTS**

→ **TORQUE**

→ **MORE STATIC EQUILIBRIUM PROBLEMS USING FORCES AND TORQUES**

→ **MOMENT OF INERTIA**

→ **ANGULAR KINEMATICS**

→ **ENERGY CONSIDERATIONS FOR ROLLING OBJECTS**

→ **ANGULAR MOMENTUM**

In Chapter 3, we discussed the fact that, if a single force is directed toward the center of mass of an extended rigid object, the result is a linear or translational acceleration in the same direction as the force. If, however, the force is not directed through the center of mass, then the result is a rotation about the center of mass. In addition, if the object is constrained by a fixed pivot point, the rotation will take place about that pivot (e.g., a hinge). These observations are summarized in Figure 7.1.

TIP

Torque produces rotation.

Figure 7.1

PARALLEL FORCES AND MOMENTS

If two forces are used, it is possible to prevent the rotation if these forces are parallel and are of suitable magnitudes. If the object is free to move in space, translational motion may result (see Figure 7.2).

parallel forces

Figure 7.2

As an example of parallel forces, consider two people on a seesaw, as shown in Figure 7.3. The two people have weights F_{g1} and F_{g2}, respectively, and are sitting distances d_1 and d_2 from the fixed pivot point (called the **fulcrum**). From our discussion about forces, we see that the tendency of each force (provided by gravity) is to cause the seesaw to rotate about the fulcrum. Force F_{g1} will tend to cause a counterclockwise rotation; force F_{g2}, a clockwise rotation. Arbitrarily, we state that a clockwise rotation is taken as being a negative, while counterclockwise is taken as positive. What factors will influence the ability of the two people to remain in "balance"; that is, what conditions must be met so that the system remains in rotational equilibrium (it is already in a state of translational equilibrium and constrained to remain that way)?

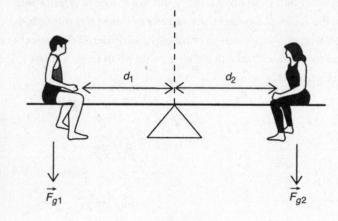

REMEMBER

In **static equilibrium**, the net torque on a system is equal to zero. In other words, the sum of the clockwise torques must be equal to the sum of the counterclockwise torques.

Figure 7.3

Through experiments, you can show that distances d_1 and d_2 (called **moment arm distances** or "lever arms") play a crucial role since we take the weights as being constant. If the two weights were equal, it should not be a surprise to learn that $d_1 = d_2$. In these examples, the weight of the seesaw is taken to be negligible (this is not a realistic scenario).

It turns out that, if F_{g2} is greater, that person must sit closer to the fulcrum. If equilibrium is to be maintained, the following condition must hold:

$$\left(F_{g1}\right)\left(d_1\right)=\left(F_{g2}\right)\left(d_2\right)$$

These are force and distance products, but they do not represent work in the translational sense because the force vectors are not causing a displacement. In rotational motion, the product of a force and a perpendicular moment arm distance (relative to a fulcrum) is called **torque**. Let us now consider some more examples of torques and equilibrium.

A 45-kg girl and a 65-kg boy are sitting on a seesaw in equilibrium. If the boy is sitting 0.7 m from the fulcrum, where is the girl sitting?

Solution

In equilibrium,

$$F_{g1}d_1 = F_{g2}d_2$$

$$(65 \text{ kg})(9.8 \text{ m/s}^2)(0.7 \text{ m}) = (45 \text{ kg})(9.8 \text{ m/s}^2)d_2$$

$$d_2 = 1.01 \text{ m}$$

TORQUE

When you tighten a bolt with a wrench, you apply a force to create a rotation or twist. This twisting action, called torque in physics, is represented by the Greek letter τ. Even though its units are newton $\cdot$ meters (N $\cdot$ m), you must not confuse it with translational work. In static equilibrium, the two conditions met are that the vector sum of all forces acting on the object equals zero and that the vector sum of all torques equals zero:

$$\Sigma \vec{F} = 0 \quad \text{and} \quad \Sigma \vec{\tau} = 0$$

When we say that $\tau = Fd$, d is the moment arm distance from the center of mass, or the pivot point, and F is the force perpendicular to the vector displacement. If the force is applied at some angle θ to the object, as seen in Figure 7.4, the component of the force, perpendicular to the vector displacement out from the pivot, is taken as the force used. Algebraically this is stated as $F_\perp = F \sin \theta$.

REMEMBER

Torque is a vector quantity, but **work** is not! The direction of the torque is taken as either positive or negative, depending on whether the rotation is counterclockwise (+) or clockwise (−).

Figure 7.4

$$\tau = Fd \sin \theta$$

Consider a light string wound around a frictionless and massless wheel, as shown in Figure 7.5. The free end of the string is attached to a 1.2-kg mass that is allowed to fall freely. The wheel has a radius of 0.25 m. What torque is produced?

$$\vec{F_g} = m\vec{g}$$

Figure 7.5

Solution

The force acting at right angles to the center of the wheel is the weight of the mass, given by

$$F_g = mg = (1.2)(9.8) = 11.76 \text{ N}$$

The radius serves as the moment arm distance, $d = 0.25$ m, so we can write

$$\tau = -Fd = (11.76)(0.25) = -2.94 \text{ N} \cdot \text{m}$$

The torque is negative since the falling weight induces a clockwise rotation in this example.

Rank the magnitude of the net torque in each of the following situations. (All forces are equal in magnitude.) The axis of rotation is indicated with a dot.

(A)

(B)

(C)

(D)

Solution

Let r be the maximum lever arm (half the length of the rotating object). Using the convention of negative torque is clockwise torque, calculate the approximate net torque in each situation:

(A) $-rF - rF = -2rF$

(B) $+rF - rF = 0$

(C) $-rF - \left(\dfrac{r}{2}\right)F = -\left(\dfrac{3}{2}\right)rF$

(D) $-\left(\dfrac{r}{2}\right)F - \left(\dfrac{r}{2}\right)F = -rF$

Rank by magnitude (absolute value) from largest torque to smallest:

$$A \to C \to D \to B$$

MORE STATIC EQUILIBRIUM PROBLEMS USING FORCES AND TORQUES

In the following example (see Figure 7.6), a hinged rod of mass M is attached to a wall with a string (and, of course, by the hinge). The string is considered massless, and makes an angle θ with the horizontal. A mass m is attached to the end of the rod, whose length is ℓ. What is the tension T in the string?

Since the system is in static equilibrium, we must write that the sum of all forces and the sum of all torques equal zero. If we choose to focus on the pivot, then all forces acting through that point do not contribute any torques since the moment arm d would be zero. The reaction force $\vec{R}$ is the response of the wall to the rod and acts at some unknown angle ϕ. Thus, we state that

$$\sum \vec{\tau} = 0 \quad \text{and} \quad \sum \vec{F} = 0$$

Situation

Free-body diagram for rod M

Figure 7.6

Since this is a static problem, both the net force and the net torques must be zero. By inspection of the diagram for hanging mass m, $T_2 = mg$. Thus, the equilibrium statement for vertical force on the rod becomes:

$$\sum F_x = R_x - T \cos \theta = 0$$
$$\sum F_y = R_y + T \sin \theta - Mg - mg = 0$$

The rod is assumed to be uniform. Therefore, its weight force is applied at the halfway point ($L/2$). By taking the left-hand side of the rod as our axis of rotation, the torque from force R is zero, leaving us with the following equilibrium statement for torques. The axis for a static problem can be taken anywhere you choose. By choosing the location where an unknown force is being applied, though, that unknown can be eliminated:

$$\sum \tau = (T \sin \theta)(L) - Mg(L/2) - mg(L) = 0$$

Note the clockwise torques are negative, while the counterclockwise torque is positive. Without additional information, the two force statements cannot be further simplified. However, the length of rod L can be eliminated from the torque equation. This allows us to solve for T in terms of the two masses:

$$T = ((M/2 + m)g)/\sin \theta$$

Usually, problems of this type will either give you the angle θ or let you determine the angle by the geometry of the situation. Once the tension $\vec{T}$ is known, one can determine the components of the unknown force $\vec{R}$ from the two force equations.

For example, if $M = 1$ kg, $m = 10$ kg, and $\theta = 30°$:

$$T = 205.8 \text{ N}$$
$$R_x = T\cos\theta = 178.2 \text{ N}$$
$$R_y = (M+m)g - T\sin\theta = 4.9 \text{ N}$$

Note that the $\vec{R}$ force from the wall is almost entirely horizontal!

It should be noted that the rod is considered to have mass and that the torque produced by the rod is its weight taken from the center of mass ($\ell/2$ in this case) and is always clockwise (negative).

MOMENT OF INERTIA

Rotational inertia is slightly more complicated than linear inertia. However, like linear inertia, rotational inertia (or an object's moment of inertia) measures an object's tendency to resist changes to its motion, which is angular motion in this case. Linear inertia is simply the total mass of an object that we imagine as being located at the object's center of mass. Rotational inertia depends not only on the mass but also on how that mass is distributed. Specifically, for a distribution of pointlike masses (or objects where the individual centers of mass are known), the moment of inertia (I) (i.e., the rotational inertia) of the entire object is given by:

$$T = \sum m_i r_i^2$$

where r is the distance from the rotational axis. The units are kg $\cdot$ m^2. The result depends not only on the shape and mass distribution of the object but also about which axis the object is being rotated. On the AP exam, if a solid object is being used, the general moment of inertia for that object will be given or easily obtained through another relationship (see below). For example, the moment of inertia of a solid rod of length L is different when rotating the object about one end of the rod as opposed to rotating it around its midpoint (see Figure 7.7).

The AP Physics 1 exam will not ask you to calculate moments of inertia; however, you must understand them and be able to use them.

about midpoint: $m \cdot L^2/12$

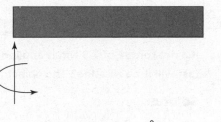

about one end: $m \cdot L^2/3$

Figure 7.7

Try oscillating a ruler between your fingers while holding it at its midpoint. Then oscillate it while holding one end. You will quickly see that rotating the ruler about its midpoint is much easier and, hence, has a smaller moment of inertia.

ANGULAR KINEMATICS

Rotational inertia takes the place of mass in all of the linear formulas in physics. An object's position in rotational motion is represented by its angular position (θ). Note that all angular values are measured in radians. (Recall the conversion 360 degrees = 2π radians = approximately 6.283 radians.) An object's angular velocity is defined analogously to linear velocity:

$$\omega = \frac{\Delta\theta}{\Delta t}(\text{rad/sec}) \text{ direction is either clockwise (CW) or counterclockwise (CCW)}$$

Likewise, angular acceleration (caused by torques) is defined kinematically by

$$\alpha = \frac{\Delta\omega}{\Delta t}\left(\text{rad/s}^2\right)$$

Having defined the basic angular quantities of torque, moment of inertia, and the kinematic variables, we can complete the analogy between the physics of linear motion and that of rotational motion. Simply rewrite any of our linear physics relationships through substitution.

Comparison of Linear and Angular/Rotational Motion

Linear	Angular/Rotational
x: position (m)	θ: angle (radians)
v: velocity (m/s)	ω: angular speed (rad/sec)
a: acceleration (m/s^2)	α: angular acceleration (rad/s^2)
$\Delta x = v_0 t + \frac{1}{2}at^2$	$\Delta\theta = \omega_0 t + \frac{1}{2}\alpha t^2$
F = force (push or pull) (newtons)	τ = torque = $|\vec{r}\,||\vec{F}|\sin\theta$
$\vec{F}_{net} = m\vec{a}$ (for constant m)	$\vec{\tau} = I\vec{\alpha}$ (for constant I)
$\vec{p} = m\vec{v}$ = momentum	$\vec{L} = I\vec{\omega}$ = angular momentum = $|\vec{r}\,||\vec{p}\,|\sin\theta$
$\vec{F}_{\|} = \Delta\vec{p}/\Delta t$	$\vec{\tau} = \Delta\vec{L}/\Delta t$
Linear KE = $\frac{1}{2}mv^2$	Rotational KE = $\frac{1}{2}I\omega^2$
Work = $|\vec{F}\,||\vec{D}|\cos\theta$	Work = $\tau \cdot \theta$

SAMPLE PROBLEM

If a net torque of 7.5 Nm is applied to an object initially at rest for 3.0 seconds, how fast will it be rotating? The object has a momentum of inertia of 15 kg m^2.

Solution

Determine the angular acceleration from the torque:

$$\vec{\tau} = I\vec{\alpha}$$

$$7.5 = 15\alpha$$

$$\alpha = 0.5 \text{ rad/s}^2$$

Use this angular acceleration in the equation of motion for angular speed:

$$\omega_f = \omega_i + \alpha_t$$

$$\omega_f = 0 + 0.5(3) = 1.5 \text{ rad/s}$$

Note that although radians (rad) are not technically a unit and are not formally required to be written down, it can be helpful to write them in to guide your thinking when solving the problem.

ENERGY CONSIDERATIONS FOR ROLLING OBJECTS

Generally speaking, objects that are rolling have both linear kinetic energy $\left(\frac{1}{2}mv^2\right.$ using the velocity of the object's center of mass) and rotational kinetic energy $\left(\frac{1}{2}I\omega^2\right)$. If the cylindrical or spherical object is rolling without slipping, then the object's rotation speed (ω) and its linear speed (v) must be coupled by the radius of the object: $v = \omega r$.

Figure 7.8 shows various objects rolling without slipping down an incline.

Figure 7.8

The mechanical energy of one of these objects is as follows:

$$mgh + \frac{1}{2}mv^2 + \frac{1}{2}I\omega^2$$

As usual with gravity problems, the mass cancels out since it is in every term. (Recall that moment of inertia, I, contains an m-term as well as an r^2-term.) Surprisingly, the rolling without slipping conditions cancels out the radius dependency in this expression as well (the r^2 in the momentum of inertia with the $\left(\frac{v}{r}\right)^2$ for ω). This means that the question of who will win the race comes down to the moments of inertia for each of the various objects. The object with the highest moment of inertia expression (not the actual value since the mass and radius cancel out) will require more of its mechanical energy for rotation. Thus, that object will have less of the total energy for the linear kinetic energy (representing the translational motion of the object). A solid sphere with a formula of $\frac{2}{5}mr^2$ is the winner in Figure 7.8!

ANGULAR MOMENTUM

Just as forces cause changes in linear momentum over time, torques cause changes in angular momentum over time. In addition, an isolated system's angular momentum is conserved just as a linear momentum is. Whereas it is relatively rare for an isolated system's mass to change, changing the moment of inertia of an isolated object is not that difficult. If the momentum of inertia changes due to internal forces, the rotational velocity must also change in order to conserve angular momentum. Observe a spinning *ice-skater* drawing in her arms, thereby decreasing the r values in her moment of inertia calculation. You will see her rotational speed increase (see Figure 7.9).

ω_1

ω_2

Arms and leg extend out

Arms and leg withdraw inward

Figure 7.9

$$I_1\omega_1 = I_2\omega_2$$

SAMPLE PROBLEM

Determine the angular momentum of Earth's orbital motion as it goes around the Sun.

Solution

Treating Earth as a point mass:

$$I = MR^2$$

where M is the mass of Earth and R is the orbital radius of the Earth about the Sun.

Next, determine the angular speed of Earth moving in a circle about the Sun:

$$\omega = v/R$$

where v is the orbital velocity of Earth.

Putting these expressions together into the angular momentum formula:

$$L = I\omega = MR^2(v/R)$$

Note that this is a general solution for the angular momentum of any point-like mass undergoing uniform circular motion!

Orbital speed of the Earth:

$$v = 2\pi R/(365 \text{ days}) \times (1 \text{ day})/(24 \text{ hours}) \times (1 \text{ hour}/3600 \text{ s}) = 2.99 \times 10^4 \text{ m}$$

$$(R = 1.5 \times 10^{11} \text{ m})$$

(Note you can get the same result by using Newton's orbital speed equation and the mass of the Sun!)

$$L_{\text{Earth}} = MvR = (6.0 \times 10^{24})(2.99 \times 10^4)(1.5 \times 10^{11}) = 2.7 \times 10^{40} \text{ kgm}^2/\text{s}$$

SUMMARY

- Torque is a vector quantity equal to the product of (force × lever arm).
- The force and lever arm are perpendicular to each other.
- In equilibrium, the sum of the clockwise torques must be equal to the sum of the counterclockwise torques.
- Newton's laws apply in rotating systems just as they do in linear systems. Everything in the rotational world has an analogous equation in the linear.
- The units for torque are N · m, but they should *not* be confused with the scalar energy unit for joules, which can also be expressed as N · m.
- Angular momentum is conserved in isolated systems.

Problem-Solving Strategies for Torque

Solving torque problems is similar to solving static equilibrium problems with Newton's laws. In fact, Newton's laws provide the first condition for static equilibrium (the vector sum of all forces equals zero). This fact means that you should once again draw a free-body diagram of the situation.

In rotational static equilibrium, the vector sum of all torques must equal zero. Remember that torque is a vector even though its units are newton · meters. We take clockwise torques as negative and counterclockwise torques as positive. Any force going through the chosen pivot point (or fulcrum) does not contribute any torques. Therefore, it is wise to choose a point that eliminates the greatest number of forces. Also, remember that only the components of forces that are perpendicular to the direction of a radius displacement from the pivot (called the moment arm) are responsible, a situation that usually involves the sine of the force's orientation angle.

Problem-Solving Strategies for Rotational Motion

Translating a rotational problem first into the more comfortable language of linear physics can be helpful. The only major difference is that whereas changes of an object's mass in the middle of a problem are rare in linear problems, changes in moment of inertia are common in rotational problems.

Multiple-Choice

1. A 45-kg girl is sitting on a seesaw 0.6 m from the balance point, as shown below. How far, on the other side, should a 60-kg boy sit so that the seesaw will remain in balance?

   (A) 0.30 m
   (B) 0.35 m
   (C) 0.40 m
   (D) 0.45 m

2. A balanced stick is shown below. The distance from the fulcrum is shown for each mass except the 10-g mass. What is the approximate position of the 10-g mass, based on the diagram?

   (A) 7 cm
   (B) 9 cm
   (C) 10 cm
   (D) 15 cm

3. A solid cylinder consisting of an outer radius r_1 and an inner radius r_2 is pivoted on a frictionless axle as shown below. A string is wound around the outer radius and is pulled to the right with a force $F_1 = 3$ N. A second string is wound around the inner radius and is pulled down with a force $F_2 = 5$ N. If $r_1 = 0.75$ m and $r_2 = 0.35$ m, what is the net torque acting on the cylinder?

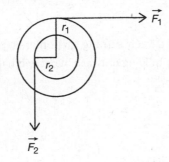

(A) 2.25 N · m

(B) −2.25 N · m

(C) 0.5 N · m

(D) −0.5 N · m

Answer questions 4 and 5 based on the following diagram. The rod is considered massless.

4. What is the net torque about an axis through point A?

(A) 16.8 N · m

(B) 15.2 N · m

(C) −5.5 N · m

(D) −7.8 N · m

5. What is the net torque about an axis through point C?

(A) 3.5 N · m

(B) 7.5 N · m

(C) −15.2 N · m

(D) 5.9 N · m

6. Compute the average angular acceleration and the angular displacement during the 2 seconds a rotating object speeds up from 0.5 rad/s to 0.7 rad/s.

(A) $\alpha = 0.1$ rad/s^2 $\Delta\theta = 0.3$ radians

(B) $\alpha = 0.2$ rad/s^2 $\Delta\theta = 0.5$ radians

(C) $\alpha = 0.1$ rad/s^2 $\Delta\theta = 1.2$ radians

(D) $\alpha = 0.2$ rad/s^2 $\Delta\theta = 1.2$ radians

7. An object with a moment of inertia I experiences a net torque of T. If the object is initially at rest, what is its angular speed after 3 seconds of this torque?

(A) $3IT$

(B) $9IT$

(C) $3I/T$

(D) $3T/I$

8. If a spinning ball of clay on top of a freely turning frictionless tabletop increases its rotational inertia by 50 percent by bulging outward, what will happen to the rotational speed of the clay and tabletop?

(A) It will be half as fast.

(B) It will be 1.5 times faster.

(C) It will be 4 times slower.

(D) It will be slower by a factor of 2/3.

Free-Response

1. A 500-N person stands 2.5 m from a wall against which a horizontal beam is attached. The beam is 6 m long and weighs 200 N (see diagram below). A cable attached to the free end of the beam makes an angle of 45° to the horizontal and is attached to the wall.

(a) Draw a free-body diagram of the beam.

(b) Determine the magnitude of the tension in the cable.

(c) Determine the reaction force that the wall exerts on the beam.

2. A uniform ladder of length ℓ and weight 100 N rests against a smooth vertical wall. The coefficient of static friction between the bottom of the ladder and the floor is 0.5.

(a) Draw a free-body diagram of this situation.

(b) Find the minimum angle, θ, that the ladder can make with the floor so that the ladder will not slip.

3. A rotating toy top (moment of inertia = 0.012 kg m^2) is rotating 8 times per second while encountering a friction torque of 0.2 Nm. For how much more time will it be rotating?

ANSWERS EXPLAINED

Multiple-Choice Problems

1. **(D)** To remain in balance, the two torques must be equal. The force on each side is given by the weight, mg. The moment arm distances are 0.6 m and x. Since the factor g will appear on both sides of the torque balance equation, we can eliminate it and write

$$(45)(0.6)=(60)x \quad \text{implies} \quad x=0.45\,\text{m}$$

2. **(C)** In this problem, again, the sum of the torques on the left must equal the sum of the torques on the right. We could convert all masses to kilograms and all distances to meters, but in the balance equation the same factors appear on both sides. Therefore, for simplicity and time efficiency, we can simply write

$$(30)(40)+(40)(20)+(20)(5)=(10)x+(50)(40)$$
$$x=10\,\text{cm}$$

3. **(D)** The net torque is given by the vector sum of all torques. $\vec{F}_2$ provides a counterclockwise positive torque, while $\vec{F}_1$ provides a clockwise negative torque. Each radius is the necessary moment arm distance. Thus, we have

$$\tau_{\text{net}}=(5)(0.35)-(3)(0.75)=-0.5\ \text{N}\cdot\text{m}$$

4. **(A)** In this problem, the net torque about point A implies that the force passing through point A does not contribute to the net torque. Also, we need the components of the remaining forces perpendicular to the beam. From the diagram, we see that the 30-N force acts counterclockwise (positive), while the 10-N force acts clockwise (negative). Thus:

$$\tau_{\text{net}}=(30)(\cos 45)(1.5)-(10)(\sin 30)(3)=16.8\ \text{N}\cdot\text{m}$$

5. **(B)** In this problem, since the pivot is now set at point C, we can eliminate the 30-N force passing through point C as a contributor to the torque. Again, we see that the 20-N force will act in a counterclockwise direction, while the 10-N force will act clockwise. Each force is 1.5 m from the pivot. We also need the component of each force perpendicular to the beam. Thus:

$$\tau_{\text{net}}=(20)(\sin 30)(1.5)-(10)(\sin 30)(1.5)=7.5\ \text{N}\cdot\text{m}$$

6. **(C)** $\alpha=\dfrac{\Delta\omega}{\Delta t}=\dfrac{0.7-0.5}{2}=0.1\,\text{rad/s/s}=0.1\,\text{rad/s}^2$

$$\Delta\theta=\omega_0 t+\frac{1}{2}\alpha t^2=0.5(2)+\frac{1}{2}(0.1)(2)^2=1+0.2=1.2\,\text{radians}$$

7. **(D)** Since $\tau=I\alpha$ and $\alpha=\dfrac{\Delta\omega}{\Delta t}$, we can combine and solve for $\Delta\omega$:

$$\Delta\omega=(T/I)\Delta t=3T/I$$

8. **(D)** Angular momentum is conserved:

$$I_i\omega_i = I_f\omega_f \qquad I_f = (3/2)I_i$$

$$I_i\omega_i = \left((3/2)I_i\right)\omega_f$$

$$\omega_f = 2/3\,\omega_i$$

Free-Response Problems

1. (a) The free-body diagram for this situation appears below:

(b) The reaction force acts at some unknown angle, θ. We use our two conditions for equilibrium:

1. The sum of all forces in the x- and y-directions must equal zero:

$$0 = R\cos\theta - T\cos 45 \quad \text{and} \quad 0 = R\sin\theta + T\sin 45 - 500\text{ N} - 200\text{ N}$$

2. The sum of all torques through contact point O must be zero, thus eliminating the reaction force R:

$$0 = -(500)(2.5) - (200)(3) + T\sin 45(6)$$

Both weights produce clockwise torques! Solving for T in the last equation, we get $T = 436.05$ N.

(c) Using the result from part (b), we can rewrite the first two equations as:

$$R\cos\theta = (436.05)\cos 45 = 308.33\text{ N} \quad \text{and} \quad R\sin\theta = 391.67\text{ N}$$

Taking the ratio of these two equations gives us $\tan\theta = 1.27$; therefore, $\theta = 51.8°$. Since we know the angle at which the reaction force acts, we can simply write

$$R\cos 51.8 = 308.33 \quad \text{implies} \quad R = 498.6\text{ N}$$

2. (a) A sketch of the situation and its free-body diagram are given below:

Situation   Free-body diagram

(b) The reaction force R is the vector resultant of the normal force N (from the floor) and the frictional force f that opposes slippage. The force P is the normal force of the vertical wall; there is no friction on the vertical wall. The weight F_g acts from the center of mass, $\ell/2$, and initiates a clockwise torque. Since we have equilibrium at this angle, we can write our equations for the conditions of static equilibrium.

1. The sum of all x- and y-forces must be equal to zero. Horizontally, friction is an opposing force, as well as the reaction force, P: $0 = f - P$. Vertically, we can write $0 = N - F_g$. Since $F_g = 100$ N, then $N = 100$ N. Now, for no slippage, $f \le \mu N$, thus, $f = (0.5)(100) = 50$ N $= P$.

2. The sum of the torques must be zero. We need the component of P perpendicular to the ladder. From the geometry, we see that this is $P \sin\theta$. Also, the component of weight perpendicular to the ladder is $F_g \cos\theta$. Thus, for the sum of all torques equals zero, we write:

$$0 = P\ell \sin\theta - F_g \left(\frac{\ell}{2}\right)\cos\theta$$

The length of the ladder is therefore irrelevant and can be canceled out. Since we know that $P = 50$ N and $F_g = 100$ N, we find that $\tan\theta = F_g/2P = 1.0$. Therefore, $\theta = 45°$.

3. First use the given torque and momentum of inertia to determine the angular deceleration:

$$\vec{\tau} = I\vec{\alpha}$$

$$0.2 = (0.012)\alpha$$

$$\alpha = 16.7 \text{ rad/s}^2$$

An object has stopped rotating when its angular speed comes to zero ($\omega_f = 0$). Use the given rotation rate to determine the initial angular speed:

$$\omega_I = 8 \text{ rot/sec} \times (2\pi \text{ rad/rot}) = 16\pi \text{ rad/sec} = 50 \cdot \text{rad/s}$$

Use the angular deceleration from above in the equation of motion for angular speed:

$$\omega_f = \omega_i + \alpha t$$

$$0 = 50 - 16.7t$$

$$t = 3.0 \text{ s}$$

Oscillatory Motion

8

KEY CONCEPTS
→ SIMPLE HARMONIC MOTION: A MASS ON A SPRING
→ SIMPLE HARMONIC MOTION: A SIMPLE PENDULUM
→ THE DYNAMICS OF SIMPLE HARMONIC MOTION

SIMPLE HARMONIC MOTION: A MASS ON A SPRING

From Hooke's law in Chapter 3, we know that a spring will become elongated by an amount directly proportional to the force applied. The force constant, k, relates to the specific amount of force (in newtons) needed to stretch or compress the spring by 1 meter:

$$\vec{F} = -k\vec{x}$$

The negative sign indicates that the force is restorative. One could easily use Hooke's law without the negative sign in the proper context.

Suppose we have a spring with a mass attached to it horizontally in such a way that the mass rests on a flat frictionless surface as shown in Figure 8.1.

Figure 8.1

> **REMEMBER**
>
> A spring will produce a restoring force if it is stretched or compressed.

If the mass M is pulled a displacement $\vec{x}$, a restoring force of $\vec{F} = -k\vec{x}$ will act on it when released. However, as the mass accelerates past its equilibrium position, its momentum will cause it to keep going, thus compressing the spring. This action will slow the mass down until the same displacement, $\vec{x}$, is reached in compression. The same restoring force will then accelerate the mass back and forth, creating oscillatory motion with a certain period T (in seconds) and frequency f (in hertz or cycles per second). The SI unit for frequency is reciprocal seconds (s^{-1}).

Now, according to Newton's second law of motion, $\vec{F} = m\vec{a}$, so when the mass was originally extended, the restoring force $\vec{F} = -k\vec{x}$ would also produce an instantaneous acceleration, given by $\vec{F} = m\vec{a}$. In other words, $\vec{a} = -(k/m)\vec{x}$. The fact that the acceleration is directly proportional to the displacement (but in the opposite direction) is characteristic of a special kind of oscillatory motion called **simple harmonic motion**. From this expression, it can be shown that the acceleration is zero at the equilibrium point (where $x = 0$).

We can build up a qualitative picture of this type of motion by considering a displacement versus time graph for this mass-spring system. Suppose we have the system at rest so that the spring is unstretched. We now pull the mass to the right a distance A (called the **amplitude**) and release the mass. This action creates a restoring force that will pull the mass to the left toward the equilibrium point. The velocity will become greater and greater, reaching a maximum as the mass passes through the equilibrium point (at which $a = 0$). The mass will then move toward the left, slowing down as it compresses the spring (since the acceleration is in the opposite direction). The mass stops momentarily when $x = -A$ (because of conservation of energy) and then accelerates again to maintain simple harmonic motion (in the absence of friction).

A graph of displacement versus time (see Chapter 2) would look like a graph of the **cosine** function since the mass is beginning at some distance from the origin. We could consider the motion in progress from the point of view of the origin, in which case the graph would be of the **sine** function (some textbooks use this format). Figure 8.2 shows the characteristics of acceleration and deceleration as the mass oscillates in a period T with a frequency f (for arbitrary units of displacement and time).

From Chapter 4, we know that the maximum energy of a compressed (or stretched) spring is given by $E = (1/2)kA^2$, for $x = A$ as in our example. Thus, the constraining points $X = \pm A$ define the limits of oscillation for the mass.

Figure 8.2

A graph of velocity versus time can be built up qualitatively in much the same way. Recall that the slope of the displacement versus time graph represents the instantaneous velocity. From Figure 8.2, you can see that, at $t = 0$, the graph is horizontal, indicating that $v = 0$. The increasingly negative slope shows that the mass is accelerating "backward" until, at $t = 1$, the line is momentarily straight, indicating maximum velocity when $x = 0$. The slope now gradually approaches zero at $t = 2$, indicating that the mass is slowing down as it approaches $x = -A$. The cycle then repeats itself, producing a graph similar to Figure 8.3.

Finally, analyzing the velocity graph with slopes, we produce an acceleration versus time graph. When $t = 0$, the line is momentarily straight with a negative slope indicating a maximum negative acceleration. When the mass crosses the equilibrium point, velocity is maximum but acceleration is momentarily zero. The acceleration (proportional to displacement) reaches a maximum once again when $x = -A$. The cycle repeats itself, producing Figure 8.4. Notice that the acceleration graph is approximately shaped like the negative of the displacement graph, as expected!

Figure 8.3

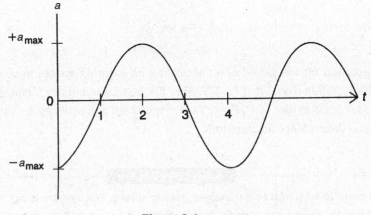

Figure 8.4

To derive the period of oscillation for the mass-spring system, we can consider another form of periodic motion already discussed: uniform circular motion. If a mass is attached to a rotating turntable and then turned onto its side, a projected shadow of the rotating mass simulates simple harmonic motion (Figure 8.5).

Figure 8.5

In this simulation, the radius r acts like the amplitude A, and the frequency and period of the rotation can be adjusted so that, when the shadow appears next to a real oscillating system, it is difficult to decide which one is actually rotating.

$$\omega = 2\pi f = \frac{2\pi}{T}$$

From our understanding of uniform circular motion, we know that the magnitude of the centripetal acceleration is given by

$$a_c = \frac{4\pi^2 r}{T^2} = \frac{v^2}{r}$$

Since the projected sideways view of this motion appears to approximate simple harmonic motion, we can let the radius r be approximated by the linear displacement x and write

$$a_x = \frac{4\pi^2 x}{T^2}$$

Since $a = (k/m)x$ in simple harmonic motion, we can now write

$$\left(\frac{k}{m}\right)x = \frac{4\pi^2 x}{T^2}$$

$$T = 2\pi\sqrt{\frac{m}{k}}$$

TIP

The period of a mass on a spring is independent of the acceleration due to gravity.

This is the equation for the period of an oscillating mass-spring system in seconds. To find the frequency of oscillation, recall that $f = 1/T$. Also, the angular frequency (velocity) is expressed as $\omega = 2\pi/T$, which implies that $\omega = \sqrt{k/m}$. The units of angular frequency are radians per second. Note there is no dependence on amplitude.

SAMPLE PROBLEM

A 0.5-kg mass is attached to a massless, elastic spring. The system is set into oscillation along a smooth, horizontal surface. If the observed period is 0.5 s, what is the value of the force constant k?

Solution

The formula for the period is

$$T = 2\pi\sqrt{\frac{m}{k}}$$

Thus,

$$k = \frac{4\pi^2 m}{T^2}$$

$$k = \frac{(4\pi^2)(0.5 \text{ kg})}{(0.5 \text{ s})^2} = 79 \text{ N/m}$$

SIMPLE HARMONIC MOTION: A SIMPLE PENDULUM

Imagine a pendulum consisting of a mass M (called a bob) and a string of length ℓ that is considered massless (Figure 8.6). The pendulum is displaced through an angle θ that is much less than 1 radian (about 57 degrees). Under these conditions, the pendulum approximates simple harmonic motion and the period of oscillation is independent of amplitude.

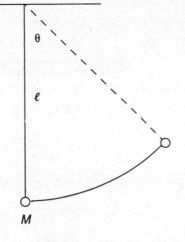

Figure 8.6

When the pendulum swings through an arc of length *s*, it appears to be following a straight path for a suitably chosen small period of time or small section of the arc. In this approximation, we can imagine the pendulum as being accelerated down an incline. The oscillations occur because gravity accelerates the pendulum back to its lowest position. Its momentum maintains the motion through that point, and then the constraining action of the string (providing a centripetal force) causes the pendulum to swing in an upward arc. Conservation of energy will bring the pendulum to the same vertical displacement (or cause it to swing through the same arc length) and then momentarily stop until gravity begins to pull it down again.

If we imagine that an incline of set angle θ is causing the acceleration, then, from our discussions of kinematics, we know that $a = -g\sin\theta$ (where the angle is measured in radians). Now, if θ is sufficiently less than 1 radian, we can write that $\sin\theta \approx \theta$ and therefore that

$$a = -g\sin\theta \approx -g\theta \approx \frac{gs}{\ell}$$

In the above equation we used the known relationship from trigonometry that, in radian measure, if *s* is the arc length and ℓ corresponds to the effective "radius" of swing, then $s = \ell\theta$.

Remember that these are only approximations, but for angles of less than 20 degrees, the approximations are fairly accurate. Since the pendulum now approximates simple harmonic motion, we know that we can find a suitable rotational motion that, when viewed in projection (as well as the pendulum swing being viewed in projection), simulates our simple harmonic motion. Therefore, in a similar fashion to that in the first section of this chapter, we can write

$$a = \frac{4\pi^2 s}{T^2} = \frac{gs}{\ell}$$

where the arc length *s* is the amplitude of swing in this case.

If we solve for the period *T*, we finally get

$$T = 2\pi\sqrt{\frac{\ell}{g}}$$

The period of a simple pendulum is independent of the mass.

Notice that period is independent of mass but is very sensitive to the local acceleration of gravity. This becomes an excellent way to independently measure the value of g in various locations.

SAMPLE PROBLEM

A simple pendulum consists of a string 0.4 m long and a 0.3-kg mass. On an unknown planet, the pendulum is set into oscillation and the observed period is 0.8 s. What is the value of g on this planet?

Solution

The period of a simple pendulum is independent of the mass and is given by

$$T = 2\pi\sqrt{\frac{\ell}{g}}$$

Thus,

$$g = \frac{4\pi^2\ell}{T^2} = \frac{(4\pi^2)(0.4 \text{ m})}{(0.8 \text{ s})^2} = 25 \text{ m/s}^2$$

The period of a system oscillating with simple harmonic motion is independent of the amplitude. In *simple harmonic motion*, the acceleration is in the opposite direction and proportional to the displacement!

THE DYNAMICS OF SIMPLE HARMONIC MOTION

From Chapter 4 we know that the work done to compress a spring is also equal to the potential energy stored in the spring. This value is expressed as $(1/2)kx^2$. If the system is set into oscillation by stretching the spring by an amount $x = A$, the maximum energy possessed by the oscillating system is given by $U = (1/2)kA^2$. As the mass oscillates, it reaches maximum velocity when it passes through $x = 0$. At that point, the potential energy of the spring is zero, and since we are treating the cases without friction, the loss of potential energy is balanced by this gain in kinetic energy:

$$KE_{\text{max}} = \frac{1}{2}mv^2{}_{\text{max}} = \frac{1}{2}kA^2$$

If we want to treat cases in which the mass is between the two extremes of $x = 0$ and $x = \pm A$, we note that the spring will still possess some elastic potential energy (no gravitational potential energy is involved since the mass is oscillating horizontally). This implies that

$$\frac{1}{2}kA^2 = \frac{1}{2}mv^2 + \frac{1}{2}kx^2$$

If we solve for velocity, we get

$$v = \pm\sqrt{\frac{k}{m}\left(A^2 - x^2\right)}$$

A mass oscillates about its equilibrium point from four different starting positions. Rank the speed of the oscillating mass from fastest speed to slowest speed. Justify your answer.

Solution

From fastest to slowest speeds, the positions are ranked as follows:

II, III, I, IV

When the spring is fully compressed or fully extended (as in position IV), the velocity drops to zero as the mass is changing direction.

When the mass is traveling through the equilibrium point (as in position II), all of the energy is kinetic; therefore, the speed is the highest.

For positions I and III, the direction of motion does not matter since the question is asking about speed, not velocity. Since position III is closer to equilibrium, it has more kinetic energy than position I, making position III the slightly faster position.

PRACTICE EXERCISES

Multiple-Choice

1. What is the length of a pendulum whose period, at the equator, is 1 s?

    (A) 0.15 m
    (B) 0.25 m
    (C) 0.30 m
    (D) 0.45 m

2. On a planet, an astronaut determines the acceleration of gravity by means of a pendulum. She observes that the 1-m-long pendulum has a period of 1.5 s. The acceleration of gravity, in meters per second squared, on the planet is

    (A) 7.5
    (B) 15.2
    (C) 10.2
    (D) 17.5

3. When a 0.05-kg mass is attached to a vertical spring, it is observed that the spring stretches 0.03 m. The system is then placed horizontally on a frictionless surface and set into simple harmonic motion. What is the period of the oscillations?

   (A) 0.75 s
   (B) 0.12 s
   (C) 0.35 s
   (D) 1.3 s

4. A mass of 0.5 kg is connected to a massless spring with a force constant k of 50 N/m. The system is oscillating on a frictionless, horizontal surface. If the amplitude of the oscillations is 2 cm, the total energy of the system is

   (A) 0.01 J
   (B) 0.1 J
   (C) 0.5 J
   (D) 0.3 J

5. A mass of 0.3 kg is connected to a massless spring with a force constant k of 20 N/m. The system oscillates horizontally on a frictionless surface with an amplitude of 4 cm. What is the velocity of the mass when it is 2 cm from its equilibrium position?

   (A) 0.28 m/s
   (B) 0.08 m/s
   (C) 0.52 m/s
   (D) 0.15 m/s

6. If the length of a simple pendulum is doubled, its period will

   (A) decrease by 2
   (B) increase by 2
   (C) decrease by $\sqrt{2}$
   (D) increase by $\sqrt{2}$

7. A 2-kg mass is oscillating horizontally on a frictionless surface when attached to a spring. The total energy of the system is observed to be 10 J. If the mass is replaced by a 4-kg mass, but the amplitude of oscillations and the spring remain the same, the total energy of the system will be

   (A) 10 J
   (B) 5 J
   (C) 20 J
   (D) 3.3 J

Free-Response

1. A mass M is attached to two springs with force constants k_1 and k_2, respectively. The mass can slide horizontally over a frictionless surface. Two arrangements, (a) and (b), for the mass and springs are shown below.

(a)

(b)

(a) Show that the period of oscillation for situation (a) is given by

$$T = 2\pi\sqrt{\frac{m(k_1 + k_2)}{k_1 k_2}}$$

(b) Show that the period of oscillation for situation (b) is given by

$$T = 2\pi\sqrt{\frac{m}{k_1 + k_2}}$$

(c) Explain whether the effective spring constant for the system has increased or decreased.

2. A mass M is attached to two light elastic strings both having length ℓ and both made of the same material. The mass is displaced vertically upward by a small displacement Δy such that equal tensions T exist in the two strings, as shown below. The mass is released and begins to oscillate up and down. Assume that the displacement is small enough so that the tensions do not change appreciably. (Ignore gravitational effects.)

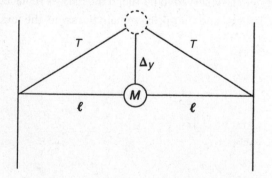

(a) Show that the restoring force on the mass can be given by (for small angles)

$$F = \frac{-2T\Delta y}{\ell}$$

(b) Derive an expression for the frequency of oscillation.

3. Explain why a simple pendulum undergoes only approximately simple harmonic motion.

4. Explain how the mass of an object can be determined in a free-fall orbit using the concept of simple harmonic motion.

5. Both oscillating springs and oscillating pendula can be used as timekeeping devices. Compare the accuracy of both of these devices as timekeeping devices here on Earth versus on the Moon where gravity is 1/6 that of Earth's.

ANSWERS EXPLAINED

Multiple-Choice Problems

1. **(D)** If we use the formula for the period of a pendulum, $T = 2\pi\sqrt{\ell/g}$, and square both sides, we can solve for the length:

$$\ell = T^2 \frac{g}{4\pi^2}$$

Using the values given in the question, we find that $\ell = 0.25$ m.

2. **(D)** Again using the formula for the period of a pendulum, this time we solve for the magnitude of the acceleration of gravity:

$$g = \frac{4\pi\ell}{T^2}$$

Using the values given in the question, we find that $g = 17.5$ m/s^2.

3. **(C)** We first find the spring constant. From $F = kx$, we see that

$$F = mg = (0.05)(9.8) = 0.49 \text{ N}$$

Thus,

$$k = \frac{F}{x} = \frac{0.49}{0.03} = 16.3 \text{ N/m}$$

Now the period is given by the formula $T = 2\pi\sqrt{m/k}$. Using the values given in the question, we find that $T = 0.35$ s.

4. **(A)** The formula for total energy is $E = (1/2)kA^2$. Using the values given in the question, we get $E = 0.01$ J.

5. **(A)** The formula for any intermediary velocity is $v = \sqrt{(k/m)(A^2 - x^2)}$. Before substituting the known values, we must change centimeters to meters: 4 cm = 0.04 m and 2 cm = 0.02 m. Then, using these values for amplitude and position, we find that $v = 0.28$ m/s.

6. **(D)** Since the period of a pendulum varies directly as the square root of the length, if the length is doubled, the period increases by $\sqrt{2}$.

7. **(A)** The maximum energy in the mass-spring system can be expressed independently of the mass; it depends on the force constant and the square of the amplitude. Thus, if the spring and the amplitude of oscillations remain the same, so will the total energy of the system (10 J).

Free-Response Problems

1. (a) When mass M is stretched a distance x, spring k_1 is stretched distance x_1 and spring k_2 is stretched distance x_2. At the point of connection, Newton's third law states that the forces must be equal and opposite. Thus, we can write

    $$k_1 x_1 = k_2 x_2$$

    Since $x = x_1 + x_2$, we have $x_2 = x - x_1$, and so

    $$k_1 x_1 = k_2 (x - x_1)$$
    $$= k_2 x - k_2 x_1$$
    $$k_1 x_1 + k_2 x_1 = k_2 x$$
    $$k_2 x = (k_1 + k_2) x_1$$
    $$x_1 = \left(\frac{k_2}{k_1 + k_2} \right) x$$

    is the expression for the displacement of k_1.

    Now, for any spring, let's say spring 1:

    $$F_1 = k_1 x_1 = \left(\frac{k_1 k_2}{k_1 + k_2} \right) x$$

    Let's call $(k_1 k_2 / k_1 + k_2) = k'$, so

    $$F_1 = k' x$$

    This is in the form of the equation for simple harmonic motion and therefore is equal to ma by Newton's second law:

    $$ma = k' x$$

    Thus,

    $$T = 2\pi \sqrt{\frac{m}{k'}} = 2\pi \sqrt{\frac{m(k_1 + k_2)}{k_1 k_2}}$$

    This is the period of oscillation for arrangement (a).

    (b) In this case, each spring is displaced the same distance as the mass, since the springs are on either side of the mass. Thus, we can write

    $$F = -(k_1 + k_2) x_1$$

    and set $k' = k_1 + k_2$. We therefore have the equation for the period of oscillation

    $$T = 2\pi \sqrt{\frac{m}{k_1 + k_2}}$$

    for arrangement (b).

    (c) Situation (a) suggests that when the springs are connected in series, the effective spring constant decreases. Situation (b) suggests that when the springs are connected in parallel, the effective spring constant increases.

2. (a) In the diagram, we can see that, since Δy is small, $\theta = \Delta y/\ell$. Now, since θ (in radians) is small, $\theta \approx \sin\theta \approx \Delta y/\ell$. From the geometry, we see that the net restoring force is given by

$$\sum F = -2T\sin\theta = \frac{-2T\Delta y}{\ell}$$

(b) Since this situation is approximating simple harmonic motion, we can write

$$F = -ky$$

$$-2T\frac{\Delta y}{\ell} = -k\Delta y$$

$$k = \frac{2T}{\ell}$$

so we must then have

$$f = \frac{1}{2\pi}\sqrt{\frac{2T}{\ell m}}$$

3. A pendulum represents only approximately simple harmonic motion since only for small angles is the acceleration proportional to the displacement (angular displacement in this case).

4. Since the period of a horizontally oscillating mass on a spring is independent of the acceleration due to gravity, the object's mass can be determined using springs and horizontal oscillations.

5. Examine our two formulas:

$$\text{Pendulum: } T = 2\pi\sqrt{\frac{\ell}{g}}$$

$$\text{Springs: } T = 2\pi\sqrt{\frac{m}{k}}$$

The pendulum will have a longer period (by a factor of $\sqrt{6}$) on the Moon since g will be six times smaller, whereas the spring's period will be unaffected. However, note that the equilibrium position for the hanging mass-on-a-spring system will be higher on the Moon as it will be stretched six times less by the weaker gravity there.

Electricity

9

KEY CONCEPTS

→ **THE NATURE OF ELECTRIC CHARGES**

→ **THE DETECTION AND MEASUREMENT OF ELECTRIC CHARGES**

→ **COULOMB'S LAW**

→ **CURRENT AND ELECTRICITY**

→ **ELECTRIC RESISTANCE**

→ **ELECTRIC POWER AND ENERGY**

→ **SERIES AND PARALLEL CIRCUITS**

→ **ELECTRICAL ENERGY IS POTENTIAL ENERGY**

THE NATURE OF ELECTRIC CHARGES

The ancient Greeks used to rub pieces of amber on wool or fur. The amber was then able to pick up small objects that were not made of metal. The Greek word for amber was *elektron* (ελεκτρον)— hence the term *electric*. The pieces of amber would retain their attractive property for some time, so the effect appears to have been **static**. The amber acted differently from magnetic ores (lodestones), naturally occurring rocks that attract only metallic objects.

In modern times, hard rubber, such as ebonite, is used with cloth or fur to dramatically demonstrate the properties of electrostatic force. If you rub an ebonite rod with cloth (charging by friction) and then bring it near a small cork sphere painted silver (called a "pith ball") and suspended on a thin thread, the pith ball will be attracted to the rod. When the ball and rod touch each other, the pith ball will be repelled. If a glass rod that has been rubbed with silk is then brought near the pith ball, the ball will be attracted to the rod. If you touch the pith ball with your hand, the pith ball will return to its normal state. This is illustrated in Figure 9.1.

In the nineteenth century, chemical experiments to explain these effects showed the presence of molecules called **ions** in solution. These ions possessed similar affinities for certain objects, such as carbon or metals, placed in the solution. These objects are called **electrodes**. The experiments confirmed the existence of two types of ions, **positive** and **negative**. The effects they produce are similar to the two types of effects produced when ebonite and glass are rubbed. Even though both substances attract small objects, these objects become **charged** oppositely when rubbed, as indicated by the behavior of the pith ball. Further, chemical experiments coupled with an atomic theory demonstrated that in solids it is the negative charges that are transferred. Additional experiments by Michael Faraday in England during the first half of the nineteenth century suggested the existence of a single, fundamental carrier of negative **electric charge**, which was later named the **electron**. The corresponding carrier of positive charge was termed the **proton**.

Figure 9.1

THE DETECTION AND MEASUREMENT OF ELECTRIC CHARGES

When ebonite is rubbed with cloth, only the part of the rod in contact with the cloth becomes charged. The charge remains localized for some time (hence the name *static*). For this reason, among others, rubber, along with plastic and glass, is called an **insulator**. A metal rod held in your hand cannot be charged statically for two reasons. First, metals are **conductors**; that is, they allow electric charges to flow through them. Second, your body is a conductor, and any charges placed in the metal rod are conducted out through you (and into the earth). This effect is called **grounding**. The silver-coated pith balls mentioned in the first section of this chapter can become statically charged because they are suspended by thread, which is an insulator. They can be used to detect the presence and sign of an electric charge, but they are not very helpful in obtaining a qualitative measurement of the magnitude of charge they possess.

An instrument that is often used for qualitative measurement is the **electroscope**. One form of electroscope consists of two "leaves" made of gold foil (Figure 9.2a). The leaves are vertical when the electroscope is uncharged. As a negatively charged rod is brought near, the leaves diverge. If we recall the hypothesis that only negative charges move in solids, we can understand that the electrons in the knob of the electroscope are repelled down to the leaves through the conducting stem. The knob becomes positively charged, as can be verified with a charged pith ball, as long as the rod is near but not touching (Figure 9.2b). If you take the rod away, the leaves will collapse as the electroscope is still neutral overall.

Upon contact, electrons are directly transferred to the knob, stem, and leaves. The whole electroscope then becomes negatively charged (Figure 9.2c). The extent to which the leaves are spread apart is an indication of how much charge is present (but only qualitatively). If you touch the electroscope, you will ground it and the leaves will collapse together.

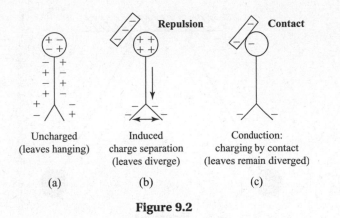

Figure 9.2

The electroscope can also be charged by **induction**. If you touch the electroscope shown in Figure 9.2b with your finger when the electroscope is brought near (see Figure 9.3a), the repelled electrons will be forced out into your body. If you remove your finger, the electroscope will be left with an overall positive charge by induction (Figure 9.3b).

Figure 9.3

Finally, we can state that electric charges, in any distribution, obey a conservation law. When we transfer charge, we always maintain a balanced accounting. Suppose we have two charged metal spheres. Sphere A has +5 elementary charges, and sphere B has a +1 elementary charge (thus, both are positively charged). The two spheres are brought into contact. Which way will charges flow? Excess charges are always spaced out as far apart as possible since they repel each other. When charges are allowed to flow, electrons do the moving even if the net charge on both objects is positive. Note that the vast majority of charge is not moving! If the two spheres are of equal size, they will each have a +3 charge after enough electrons move from the +1 sphere to the +3 sphere in this case. If one object is bigger than the other, the larger object will wind up with more of the net excess charge after those excess charges have all spread out as evenly as possible around the outer surfaces of the combined object.

COULOMB'S LAW

From the first two sections of this chapter, we can conclude that like charges repel (see Figure 9.4) each other, while unlike charges attract. The electrostatic force between two charged objects can act through space and even a vacuum. This property makes electrostatic force similar to the force of gravity.

Figure 9.4

THINK ABOUT IT

Coulomb's law is an *inverse square* law that is very similar to Newton's law of gravitation. However, the electrostatic force can be neutralized while gravity is always present. Additionally, with the electrostatic force, there can be repulsion as well as attraction.

In the late eighteenth century, the nature of the electrostatic force was studied by French scientist Charles Coulomb. He discovered that the force between two point charges (designated as q_1 and q_2), separated by a distance r, experienced a mutual force along a line connecting the charges that varied directly as the product of the charges and inversely as the square of the distance between them. This law, known as **Coulomb's law**, is, like the law of gravity, an inverse square law acting on matter at a distance. Mathematically, Coulomb's law can be written as

$$F = \frac{kq_1 q_2}{r^2}$$

The constant k has the value $8.9875 \times 10^9 \approx 9 \times 10^9 \, \text{N} \cdot \text{m}^2/\text{C}^2$.

Coulomb's law, like Newton's law of universal gravitation, is a vector equation. The direction of the force is along a radial vector connecting the two point sources. It should be noted that Coulomb's law applies only to one pair of point sources (or to sources that can be treated as point sources, such as charged spheres). If we have a distribution of point charges, the net force on one charge is the vector sum of all the other electrostatic forces. This aspect of force addition is sometimes termed **superposition**.

SAMPLE PROBLEM

Calculate the static electric force between a $+6.0 \times 10^{-6}$ C charge and a -3.0×10^{-6} C charge separated by 0.1 m. Is this an attractive or repulsive force?

Solution

We use Coulomb's law:

$$F = \frac{kq_1 q_2}{r^2}$$

$$F = \frac{(9 \times 10^9 \, \text{N} \cdot \text{m}^2/\text{C}^2)(+6.0 \times 10^{-6}\,\text{C})(-3.0 \times 10^{-6}\,\text{C})}{(0.1\,\text{m})^2} = -16.2\,\text{N}$$

The negative sign (opposite charges) indicates that the force is attractive.

Three charges are near each other as shown below (all distances are center to center). Calculate the net electrostatic force (magnitude and direction) on the bottom left charge below:

Solution

First, determine the direction of each force:

Let F_1 be the electric force of 175 µC charge on the −512 µC charge. It will be straight upwards.

Let F_2 be the electrical force of 451 µC charge on the −512 µC charge. It will be straight to the right.

Next, determine the magnitudes:

$$F_1 = Kq_1q_2/r^2 = (9 \times 10^9)(512)(175)(10^{-12})/(1.75)^2 = 263 \text{ N}$$

$$F_2 = Kq_1q_3/r^2 = (9 \times 10^9)(512)(451)(10^{-12})/(1.55)^2 = 865 \text{ N}$$

	x-component	y-component
F_1	0	263
F_2	865	0
F_{NET}	865	263

Magnitude = $(865^2 + 263^2)^{1/2} = 904 \text{ N}$

Angle = $\tan^{-1}(263/865) = 16.9°$

CURRENT AND ELECTRICITY

Previously in this chapter, we observed that, if two points have a potential difference between them, and they are connected with a conductor, negative charges will flow from a higher concentration to a lower one. This aspect of charge flow is very similar to the flow of water in a pipe due to a pressure difference.

Moving electric charges are referred to as electric **current**, which measures the amount of charge passing a given point every second. The units of measurement are coulombs per second (C/s). These are defined as an ampere or "amp." Algebraically, we designate current by the capital letter I and state that

$$I = \frac{\Delta Q}{\Delta t}$$

In electricity, it is the battery that supplies the potential difference needed to maintain a continuous flow of charge. In the nineteenth century, physicists thought that this potential difference was an electric force, called the **electromotive force** (emf), that pushed an electric fluid through a conductor. Today, we know that an emf is not a force, but rather a potential difference measured in volts. Do not be confused by the designation emf in the course of reviewing or of solving problems!

From chemistry, recall that a battery uses the action of acids and bases on different metals to free electrons and maintain a potential difference. In the process, two terminals, designated positive and negative, are created. When a conducting wire is attached and looped around to the other end, a complete circle of wire (a **circuit**) is produced, allowing for the continuous flow of charge. The battery acts like an elevator, raising electrons from the positive side up to the negative side using chemical reactions (see Figure 9.5). These electrons can then do work by transforming their electric potential energy into other forms of energy. This work is the electricity with which we have become so familiar in our modern world.

Figure 9.5

The diagram in Figure 9.5 shows a simple electric circuit. The direction of the conventional current, like electric fields, is from the positive terminal. To maintain a universal acceptance of concepts and ideas (recall our earlier discussions of concepts and labels), schematic representations for electrical devices were developed and accepted by physicists and electricians worldwide. These schematics are used when drawing or diagramming an electric circuit, and it is important that you be able to interpret and draw them to fully understand this topic. In Figure 9.6, schematics for some of the most frequently encountered electrical devices are presented.

Figure 9.6 Electrical Schematic Diagrams

The simple circuit shown in Figure 9.5 can now be diagrammed schematically as in Figure 9.7.

Figure 9.7

A switch has been added to the schematic; of course the charge would not flow unless the switch was closed. An open switch stops the flow by breaking the circuit.

In Figure 9.6, three schematics appear that we have not yet discussed. A **resistor** is a device whose function is to use up voltage. We will investigate resistors in more detail in the next section.

An **ammeter** is a device that measures the current. You can locate the water meter in your house or apartment building and notice that it is placed within the flow line. The reason is that the meter must measure the flow of water per second through a given point. An ammeter is placed within an electric circuit in much the same way. This is referred to as a "series connection," and it maintains the singular nature of the circuit. In practical terms, you can imagine cutting a wire in Figure 9.7, and hooking up the bare leads to the two terminals of the ammeter. Ammeters have very low resistance.

A **voltmeter** is a device that measures the potential difference, or voltage, between two points. Unlike an ammeter, the voltmeter cannot be placed within the circuit since it will effectively be con-nected to only one point. A voltmeter is therefore attached in a "parallel connection," creating a second circuit through which only a small amount of current flows to operate the voltmeter. In Figure 9.8, the simple circuit is redrawn with the ammeter and voltmeter placed. Volt-meters have very high resistance.

A volt is the unit of electric potential: 1 volt = 1 joule/coulomb. A volt represents the amount of energy per charge relative to either a predetermined "ground" ($V = 0$) or between two points. Voltage, potential, potential difference, electromotive force (emf), voltage drop, and electric potential are all terms used for this important concept in electricity.

Figure 9.8

For simple circuits, there will be no observable difference in readings if the ammeter and voltmeter are moved to different locations. However, there is a slight difference in the emf across the terminals of the battery when the switch is closed versus when it is open. The first emf reflects the work done by the battery not only to move charge through the circuit but also to move charge across the terminals of the battery. This is sometimes referred to as the **terminal emf** or **internal emf**.

> When a battery dies, it does not run out of volts or charge; it runs out of energy. The battery's voltage rating is a measure of how much energy the battery will give up to each charge that passes through it. The battery will continue to do so until it runs out of energy. When you recharge a battery, you are replenishing its energy!

ELECTRIC RESISTANCE

In Figure 9.8, a simple electric circuit is illustrated with measuring devices for voltage and current. If a lightbulb is left on for a long time, two observations can be made. First, the bulb gets hot because of the action of the electricity in the filament. The light produced by the bulb is caused by the heat of the filament. Second, the current in the ammeter will begin to decrease.

These two observations are linked to the idea of electrical resistance. The interaction of flowing electrons and the molecules of a wire (or bulb filament) creates an electrical **resistance**. This resistance is temperature dependent since, from the kinetic theory, an increase in temperature will increase the molecular activity and therefore interfere to a greater extent with the flow of current. Electrical potential energy is being converted to thermal energy.

In the case of a lightbulb, it is this resistance that is desired in order for the bulb to do its job. Resistance along a wire or in a battery, however, is unwanted and must be minimized. In more complicated circuits, a change in current flow is required to protect devices, and so special resistors are manufactured that are small enough to easily fit into a circuit. While resistance is the opposite of conductance, we do not want to use insulators as resistors since insulators will stop the flow altogether. Therefore, a range of materials, indexed by "resistivity," is catalogued in electrical handbooks to assist scientists and electricians in choosing the proper resistor for a given situation.

If the temperature can be maintained at a constant level, a simple relationship between the voltage and the current in a circuit is revealed: as the voltage is increased, a greater flow of current is observed. This direct relationship was investigated theoretically by a German physicist named Georg Ohm and is called **Ohm's law** (see Figure 9.9).

Figure 9.9 Ohm's law

Ohm's law states that in a circuit at constant temperature the ratio of the voltage to the current remains constant. The slope of the line in Figure 9.9 represents the resistance of the circuit, which is measured in units of ohms (Ω). Algebraically, we can write Ohm's law as

$$R = \frac{V}{I} \quad \text{or} \quad V = IR$$

Not all conductors obey Ohm's law. Lightbulbs, semiconductors, and liquid conductors, for example, do not.

Other factors also affect the resistance of a conductor. We have already discussed the effects of temperature and material type (resistivity). Resistivity is designated by the Greek letter ρ (rho). In a wire, electrons try to move through, interacting with the molecules that make up the wire. If the wire has a small cross-sectional area, the chances of interacting with a bound molecule increase, and thus the resistance of the wire increases. If the length of the wire is increased, the greater duration of interaction time will also increase the resistance of the wire. In summary, these resistance factors all contribute to the overall resistance of the circuit.

Algebraically, we may write these relationships (at constant temperature) in the following form:

$$R = \frac{\rho L}{A}$$

where L is the length in meters and A is the cross-sectional area in units of m^2. The resistivity (ρ) is measured in units of ohm · meters ($\Omega \cdot$ m) and is usually rated at 20 degrees Celsius.

In Table 9.1, the resistivities of various materials are presented. Since the ohm is a standard (SI) unit, be sure that all lengths are in meters and areas are in square meters.

Table 9.1. Resistivities of Selected Materials at 20°C

Substance	Resistivity, ρ ($\Omega \cdot$ m)
Aluminum	2.83×10^{-8}
Carbon	3.5×10^{-5}
Copper	1.69×10^{-8}
Gold	2.44×10^{-8}
Nichrome	100×10^{-8}
Platinum	10.4×10^{-8}
Silver	1.59×10^{-8}
Tungsten	5.33×10^{-8}

SAMPLE PROBLEM

Copper wire is being used in a circuit. The wire is 1.2 m long and has a cross-sectional area of 1.2×10^{-8} m^2 at a constant temperature of 20°C.

(a) Calculate the resistance of the wire.

(b) If the wire is connected to a 10-V battery, what current will flow through it?

Solution

(a) We use

$$R = \rho L/A \text{ where in this case, } \rho \text{ is the resistivity at 20°C}$$

$$R = \frac{(1.69 \times 10^{-8} \ \Omega \cdot \text{m})(1.2 \text{ m})}{1.2 \times 10^{-8} \text{ m}^2} = 1.69 \ \Omega$$

(b) Now, we use Ohm's law

$$I = \frac{V}{R} = \frac{10 \text{ V}}{1.69 \ \Omega} = 5.9 \text{ A}$$

ELECTRIC POWER AND ENERGY

Electrical energy can be used to produce light and heat. Electricity can do work to turn a motor. Having measured the voltage and current in a circuit, we can determine the amount of power and energy being produced in the following way. The units of voltage (potential difference) are joules per coulomb (J/C) and are a measure of the energy supplied to each coulomb of charge flowing in the circuit. The current, I, measures the total number of coulombs per second flowing at any given time. The product of the voltage and the current (*VI*) is therefore a measure of the total power produced since the units will be joules per second (watts):

$$P = VI$$

The electrical energy expended in a given time *t* (in seconds) is simply the product of the power and the time:

$$\text{Energy} = Pt = VIt$$

Unit analysis

$$\left(\frac{J}{C}\right) \times \left(\frac{C}{s}\right) = \frac{J}{s} = \text{watt}$$

Derivation (recalling that change in energy for electricity is *qV*)

$$P = \frac{\text{work}}{\text{time}}$$
$$= \frac{qV}{t}$$
$$= \left(\frac{q}{t}\right)V$$
$$= IV$$

SAMPLE PROBLEM

How much energy is used by a 10-Ω resistor connected to a 24-V battery for 30 minutes?

Solution

We know that the energy is given by $E = VIt = (V^2/R)t$, where *t* must be in seconds.

Thus,

$$E = \frac{(10\,\text{V})^2(1,800\,\text{s})}{10\,\text{W}} = 103,680\,\text{J}$$

SERIES AND PARALLEL CIRCUITS

Two very important conservation laws underlie all of the circuit analysis that this chapter is about to explore: conservation of charge and conservation of energy. Gustav Kirchhoff first applied these to circuits and gave us the following rules for circuits.

1. **The junction rule:** The total current coming into a junction must equal the total current leaving the junction. (Charge must be conserved.)

2. **The loop rule:** The total voltage drops and gains must total to zero as you travel around any closed loop of a circuit. (Energy must be conserved.) Traveling across a resistor with the current is a voltage drop, against the current a gain. Traveling across a battery from negative to positive is a voltage gain and from positive to negative is a voltage drop.

What is the current in the top path of the following section of a circuit?

Solution

Since the current coming into the junction on the left is 9 amps, a total of 9 amps must be leaving. Since the other two branches are carrying away 8 amps total, 1 amp is left for the missing pathway.

How much voltage is being used by the resistor R_1 in this picture?

Solution

If we trace a loop starting at the bottom right-hand corner and go counterclockwise, we encounter the following voltage changes, which must sum to zero:

$$+9V + (R_1)I - 5V - 2V = 0$$

Solving for $(R_1)I$ gives us $-2V$, representing the voltage drops across the resistor R_1 by 2 volts. This means current is going from right to left in this resistor as we traced through this resistor from right to left to obtain the decrease in voltage result. (Recall that current goes from high voltage to low voltage.)

Series Circuits

A series circuit consists of two or more resistors sequentially placed within one circuit. More generally, elements on the same path within a more complicated circuit are "in series with each other." An example is seen in Figure 9.10.

Figure 9.10 Series Circuit

We need to ask two questions. First, what is the effect of adding more resistors in series to the overall resistance in the circuit? Second, what effect does adding the resistors in series have on the current flowing in the circuit?

One way to think about this circuit is to imagine a series of doors, one after the other. As people exit through one door, they must wait to open another. The result is to drain energy out of the system and decrease the number of people per second leaving the room. In an electric circuit, adding more resistors in series decreases the current (with the same voltage) by increasing the resistance of the circuit.

Algebraically, all of these observations can be summarized as follows. In resistors R_1, R_2, and R_3, we have currents I_1, I_2, and I_3. Each of these three currents is equal to each other current and to the circuit current I:

$$I = I_1 = I_2 = I_3$$

However, the voltage across each of these resistors is less than the source voltage V. If all three resistors were equal, each voltage would be equal to one-third of the total source voltage. In any case, we have

$$V = V_1 + V_2 + V_3$$

Using Ohm's law ($V = IR$), we can rewrite this expression as

$$IR = I_1 R_1 + I_2 R_2 + I_3 R_3$$

However, the three currents are equal, so they can be canceled out of the expression. This leaves us with

$$R = R_1 + R_2 + R_3$$

In other words, when resistors are added in series, the total resistance of the circuit increases as the sum of all the resistances. This fact explains why the current decreases as more resistors are added. In our example, the total resistance of the circuit is

$$1\,\Omega + 2\,\Omega + 3\,\Omega = 6\,\Omega$$

Since the source voltage is 12 V, Ohm's law provides for a circuit current of $12/6 = 2$ A. The voltage across each resistor can now be determined from Ohm's law since each gets the same current of 2 A. Thus,

$$V_1 = (1)(2) = 2\,\text{V}; \quad V_2 = (2)(2) = 4\,\text{V}; \quad V_3 = (3)(2) = 6\,\text{V}$$

The voltages in a series circuit add up to the total, and not unexpectedly we have

$$2\,\text{V} + 4\,\text{V} + 6\,\text{V} = 12\,\text{V}$$

It should be noted that, when batteries are connected in series (positive to negative), the effective voltage increases additively as well.

Parallel Circuits

A parallel circuit consists of multiple pathways connected from one point to another, all experiencing the same potential difference. An example of a parallel circuit is seen in Figure 9.11.

Figure 9.11 Parallel Circuit

In this circuit, a branch point is reached in which the current I is split into I_1 and I_2. Since each resistor is connected to a common potential point, experiments verify that the voltage across each resistor equals the source voltage V. Thus, in this circuit it is the current that is shared; the voltage is the same. Another feature of the parallel circuit is the availability of alternative paths. If one part of the circuit is broken, current can flow through the other path. Although each branch current is less than the total circuit current I, the effect of adding resistors in parallel is to increase the effective circuit current by decreasing the circuit resistance R.

To understand this effect further, imagine a set of doors placed next to each other along a wall in a room. Unlike the series connection (in which the doors are placed one after the other), the parallel-circuit analogy involves placing the doors next to each other. Even though each door will have fewer people per second going through it at any given time, the overall effect is to allow, in total, more people to exit from the room. This is analogous to reducing the circuit resistance and increasing the circuit current (at the same voltage).

Algebraically, we can express these observations as follows. We have two resistors, R_1 and R_2, with currents I_1 and I_2. Voltmeters placed across the two resistors would indicate voltages V_1 and V_2, which would be essentially equal to the source voltage V; that is, in this example,

$$V = V_1 = V_2$$

Ammeters placed in the circuit would reveal that the circuit current I is equal to the sum of the branch currents I_1 and I_2; that is,

$$I = I_1 + I_2$$

Using Ohm's law, we see that

$$\frac{V}{R} = \frac{V_1}{R_1} + \frac{V_2}{R_2}$$

Since all voltages are equal, they can be canceled from the expression, leaving us with the following expression for resistors in parallel:

$$\frac{1}{R} = \frac{1}{R_1} + \frac{1}{R_2}$$

This expression indicates that the total resistance is determined "reciprocally," which in effect reduces the total resistance of the circuit as more resistors are added in parallel. If, for example, $R_1 = 10\ \Omega$ and $R_2 = 10\ \Omega$, the total resistance is $R = 5\ \Omega$ (in parallel)! It should be noted that connecting batteries in parallel (positive to positive and negative to negative) has no effect on the overall voltage of the combination.

SAMPLE PROBLEM

A 20-Ω resistor and a 5-Ω resistor are connected in parallel. If a 16-V battery is used, calculate the equivalent resistance of the circuit, the circuit current, and the amount of current flowing through each resistor.

Solution

We know that the equivalent resistance is given by

$$\frac{1}{R_{eq}} = \frac{1}{R_1} + \frac{1}{R_2} = \frac{1}{20\ \Omega} + \frac{1}{5\ \Omega}$$

In this case, it is easily seen that $R_{eq} = 4\ \Omega$.

Thus, using Ohm's law,

$$I = \frac{V}{R_{eq}} = \frac{16\ V}{4\ \Omega} = 4\ A$$

To find the current in each branch, we recall that the voltage drop across each resistor is the same as the source voltage. In this case:

$$I_{20} = \frac{16\ V}{20\ \Omega} = 0.8\ A$$

$$I_5 = \frac{16\ V}{5\ \Omega} = 3.2\ A$$

Notice that the total current is equal to 4 A as expected.

In Figure 9.12, a circuit that consists of resistors in series and parallel is presented. The key to reducing such a circuit is to decide whether it is, overall, a series or parallel circuit. In Figure 9.12, an 8-Ω resistor is placed in series with a parallel branch containing two 4-Ω resistors. The problem is to reduce the circuit to only one resistor and then to determine the circuit current, the voltage reading in the voltmeter, and the current reading in the meter.

Figure 9.12

Solution

To reduce the circuit and find the circuit resistance R, we must first reduce the parallel branch. Thus, if R_e is the equivalent resistance in the branch, then

$$\frac{1}{R_e} = \frac{1}{4} + \frac{1}{4}$$

This implies that $R_e = 2\ \Omega$ of resistance. The circuit can now be thought of as a series circuit between a 2-Ω resistor (replacing the portion of the circuit measured by the voltmeter) and the existing 8-Ω resistor. Thus,

$$R = 2\ \Omega + 8\ \Omega = 10\ \Omega$$

Since the circuit voltage is 20 V, Ohm's law states that the circuit current is

$$I = \frac{V}{R} = \frac{20}{10} = 2\,\text{A}$$

In a series circuit, the voltage drop across each resistor is shared proportionally since the same current flows through each resistor. Thus, for the branch, the voltmeter will read

$$V = IR_e = (2)(2) = 4\,\text{V}$$

To find the current in each part of the parallel branch, we recall that in a parallel circuit the voltage is the same across each resistor. Thus, since the two resistors are equal, each will get half of the circuit current; thus, each current and the reading on the ammeter is 1 A. Note how we used Kirchhoff's rule at the junction to halve the current.

TIP

Expect questions about simple circuits on the AP Physics 1 exam to be limited to one parallel path and one ideal battery in circuits.

ELECTRICAL ENERGY IS POTENTIAL ENERGY

Many students of physics can get confused about the basic fact that the current going into a resistor is the same as the current coming out of the resistor. How can this be when the electrons have lost energy as the resistor heated up or gave off light? The conflict arises because many students are imagining the energy of the moving electrons as being kinetic energy. The energy in circuits, however, is electrical potential energy! The electrical potential energy is carried in the electric and magnetic fields associated with the moving charges. The electrons going into the resistor are spaced differently going into the resistor than they are when they are coming out. Even though the number of electrons passing per second, and hence their average speed (drift velocity), is the same before and after, the fields associated with their relative positions are completely different. Recall that all potential energies are energies of physical relationship between two or more interacting objects. Just as 5 fully extended rubber bands moving past you at 5 mph have more total energy than the same 5 slack rubber bands moving past you at the same speed, the slightly closer-spaced electrons before they enter the resistor have more energy than the slightly farther apart electrons leaving the resistor.

The purest example of this field energy for electricity is in electromagnetic radiation (or light). The light travels without any charges (or mass for that matter). Yet the light is demonstrably carrying energy in its fields as it goes from one place to another. Parallels can be made with other potential energies. The gravitational potential of a rock held above the ground is not actually stored within the rock itself but, rather, in the gravitational field between the rock and Earth. The elastic energy of a stretched spring does not belong to the block at the end of the spring but is in the stretched or compressed spring itself.

SUMMARY

- There are two kinds of electric charges: positive and negative.
- Electrons carry a negative charge.
- Protons carry a positive charge.
- Like charges repel, while unlike charges attract.
- An electroscope can be used to detect the presence of static charges.
- Objects become charged through the transfer of electrons.
- Coulomb's law describes the nature of the force between two static charges. It states that the force of either attraction (–) or repulsion (+) is proportional to the product of the charges and inversely proportional to the square of the distance between them. This is similar to Newton's law of gravitation (see Chapter 5).
- The electrical potential difference is defined to be equal to the work done per unit charge (in units of joules per coulomb or volts).
- Electric current is a measure of the flow of charge in units of amperes (1 A = 1 C/s).
- The conventional current is based on the direction of positive charge flow.
- Ohm's law states that at constant temperature the ratio of voltage and current is a constant in a conductor ($V/I = R$).
- Electrical resistance is based on the material used ("resistivity"), the length, and the cross-sectional area at constant temperature ($R = \rho L/A$).
- Ammeters measure current and are placed "in series" within the circuit.
- Voltmeters measure potential difference (voltage) and are placed in parallel (across segments of the circuit).
- A closed path and a source of potential difference are needed for a simple circuit.
- Resistors connected in series have an equivalent resistance equal to their numerical sum ($R_{eq} = R_1 + R_2$) and carry the same current through each.
- Resistors connected in parallel have an equivalent resistance equal to their reciprocal sum [$1/R_{eq} = (1/R_1) + (1/R_2)$] and experience the same potential difference across each.
- Kirchhoff's rules describe the flow of current in circuit branches and the changes in voltage around loops.

Problem-Solving Strategies for Electric Circuits

Several techniques for solving electric circuit problems have been discussed in this chapter. For resistances with one source of emf, we use the techniques of series and parallel circuits. Ohm's law plays a big role in measuring the potential drops across resistors and the determination of currents within the circuit.

When working with a combination circuit, try to reduce all subbranches first, as well as any series combinations. The goal is to be left with only one equivalent resistor and then use Ohm's law to identify the various missing quantities. Drawing a sketch of the circuit (if one is not already provided) is essential. Also, the direction of current is taken from the positive terminal of a battery. This is opposite to the way that electrons actually move!

Multiple-Choice

1. An insulated metal sphere, A, is charged to a value of $+Q$ elementary charges. It is then touched and separated from an identical but neutral insulated metal sphere, B. This second sphere is then touched to a third identical and neutral insulated metal sphere C. Finally, spheres A and C are touched and then separated. Which of the following represents the distribution of charge on each sphere, respectively, after the above process has been completed?

    (A) $Q/3$, $Q/3$, $Q/3$
    (B) $Q/4$, $Q/2$, $Q/4$
    (C) $3Q/8$, $Q/4$, $3Q/8$
    (D) $3Q/8$, $Q/2$, $Q/4$

2. Electrical appliances are usually grounded in order to

    (A) maintain a balanced charge distribution
    (B) prevent the buildup of heat on them
    (C) run properly using household electricity
    (D) prevent the buildup of static charges on their surfaces

3. Two point charges experience a force of repulsion equal to 3×10^{-4} N when separated by 0.4 m. If one of the charges is 5×10^{-4} C, what is the magnitude of the other charge?

    (A) 1.07×10^{-11} C
    (B) 2.7×10^{-11} C
    (C) 3.2×10^{-9} C
    (D) 5×10^{-4} C

4. How many electrons are moving through a current of 2 A for 2 s?

    (A) 3.2×10^{-19}
    (B) 6.28×10^{18}
    (C) 4
    (D) 2.5×10^{19}

5. How much electrical energy is generated by a 100-W lightbulb turned on for 5 minutes?

    (A) 20 J
    (B) 500 J
    (C) 3,000 J
    (D) 30,000 J

6. What is the current in a 5-Ω resistor due to a potential difference of 20 V?

    (A) 25 A
    (B) 4 A
    (C) 0.20 A
    (D) 5 A

7. A 20-Ω resistor has 10 A of current in it. The power generated is

(A) 2,000 W

(B) 2 W

(C) 200 W

(D) 4,000 W

8. What is the value of resistor R in the circuit shown below?

(A) 20 Ω

(B) 2 Ω

(C) 4 Ω

(D) 12 Ω

Answer questions 9 and 10 based on the circuit below:

9. If the current in the circuit above is 4 A, then the power generated by resistor R is

(A) 6 W

(B) 18 W

(C) 24 W

(D) 7 W

10. What is the value of resistor R in the circuit above?

(A) 4 Ω

(B) 3 Ω

(C) 5 Ω

(D) 6 Ω

11. A 5-Ω and a 10-Ω resistor are connected in series with one source of emf of negligible internal resistance. If the energy produced in the 5-Ω resistor is X, then the energy produced in the 10-Ω resistor is

(A) X

(B) $2X$

(C) $X/2$

(D) $X/4$

12. Four equivalent lightbulbs are arranged in a circuit as shown below. Which bulb will be the brightest?

(A) A

(B) B

(C) C

(D) D

Free-Response

1. Four identical point charges ($Q = +3$ µC) are arranged in the form of a rectangle as shown below. What are the magnitude and the direction of the net electric force on the charge in the lower left corner due to the other three charges?

2. An electrostatically charged rod attracts a small, suspended sphere. What conclusions can you make about the charge on the suspended sphere?

3. Why is it safe to be inside a car during a lightning storm?

4. A 60-Ω resistor is made by winding platinum wire into a coil at 20°C. If the platinum wire has a diameter of 0.10 mm, what length of wire is needed?

5. Based on the circuit shown below:

(a) Determine the equivalent resistance of the circuit.
(b) Determine the value of the circuit current I.
(c) Determine the value of the reading of ammeter A.
(d) Determine the value of the reading of voltmeter V.

6. Show that for two resistors in parallel, the equivalent resistance can be written as

$$R_{eq} = \frac{R_1 R_2}{R_1 + R_2}$$

7. Explain why connecting two batteries in parallel does not increase the effective emf.

8. You have five 2-Ω resistors. Explain how you can combine them in a circuit to have equivalent resistances of 1, 3, and 5 Ω.

9. A laboratory experiment is performed using lightbulbs as resistors. After a while, the current readings begin to decline. What can account for this?

ANSWERS EXPLAINED

Multiple-Choice Problems

1. **(C)** We use the conservation of charge to determine the answer. Sphere A has $+Q$ charge, and B has zero. When these spheres are touched and separated, each will have one-half of the total charge, which was $+Q$. Thus, each now has $+Q/2$.

    When B is touched with C, which also has zero charge, we again distribute the charge evenly by averaging. Thus, B and C will now each have $+Q/4$, while A still has $+Q/2$.

    Finally, when A and C are touched, we take the average of $+Q/4$ and $+Q/2$, which is $+3Q/8$.

    The final distribution is therefore $3Q/8$, $Q/4$, $3Q/8$.

2. **(D)** Electrical appliances are grounded to prevent the buildup of static charges on their surfaces.

3. **(A)** We use Coulomb's law to calculate the answer. Everything is given except one charge. The working equation therefore looks like this:

$$3\times10^{-4} = \frac{(9\times10^9)(5\times10^{-4})q}{(0.4)^2}$$
$$q = 1.07\times10^{-11}\,\text{C}$$

4. **(D)** By definition, 1 A of current measures 1 C of charge passing through a circuit each second. One coulomb is the charge equivalent of 6.28×10^{18} electrons. In this question, we have 2 A for 2 s. This is equivalent to 4 C of charge or 2.5×10^{19} electrons.

5. **(D)** Energy is equal to the product of power and time. The time must be in seconds; 5 min = 300 s. Thus,

$$\text{Energy} = (100)(300) = 30{,}000\,\text{J}$$

6. **(B)** Using Ohm's law, we have

$$I = \frac{V}{R} = \frac{20}{5} = 4\,\text{A}$$

7. **(A)** We know that $P = VI$, and using Ohm's law $(V = IR)$, we get

$$P = I^2R = (100)(20) = 2{,}000\,\text{W}$$

8. **(C)** Using Ohm's law, we can determine the equivalent resistance of the circuit: $36/3 = 12\,\Omega$. The two resistors are in series, so $12 = 8 + R$. Thus, $R = 4\,\Omega$.

9. **(A)** In a parallel circuit, all resistors have the same potential difference across them. The branch currents must add up to the source current (4 A, in this case). Thus, using Ohm's law, we can see that the current in the 2-Ω resistor is $I = 6/2 = 3$ A. Therefore, the current in resistor R must be equal to 1 A. Since $P = VI$, the power in resistor R is

$$P = (6)(1) = 6\,\text{W}$$

10. **(D)** Using Ohm's law, we have,

$$R = \frac{V}{I} = \frac{6}{1} = 6\,\Omega$$

11. **(B)** In a series circuit, the two resistors have the same current but proportionally shared potential differences. Thus, since the 10-Ω resistor has twice the resistance of the 5-Ω resistor in series, it has twice the potential difference. Therefore, with the currents equal, the 10-Ω resistor will generate twice as much energy, that is, $2X$, when the circuit is on.

12. **(D)** The fact that all the resistors (bulbs) are equal means that the parallel branch will reduce to an equivalent resistance less than that of resistor D. The remaining series circuit will produce a greater potential difference across D than across that equivalent resistor. With the currents equal at that point, resistor D will generate more power than the equivalent resistance. Thus, with the current split for the actual parallel part, even less energy will be available for resistors A, B, and C. Bulb D will be the brightest.

Free-Response Problems

1. Since all the charges are positive, the lower-left-corner charge will experience mutual repulsions from the other three. A diagram showing these forces appears below:

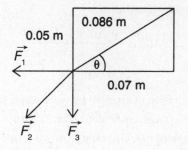

The magnitude of each force is given by Coulomb's law. The diagonal distance was determined using the Pythagorean theorem. The three force magnitudes are therefore calculated to be:

$$F_1 = \frac{(9 \times 10^9)(3 \times 10^{-6})(3 \times 10^{-6})}{(0.07)^2} = 16.53 \text{ N}$$

$$F_2 = \frac{(9 \times 10^9)(3 \times 10^{-6})(3 \times 10^{-6})}{(0.086)^2} = 10.95 \text{ N}$$

$$F_3 = \frac{(9 \times 10^9)(3 \times 10^{-6})(3 \times 10^{-6})}{(0.05)^2} = 32.4 \text{ N}$$

Now, $\vec{F}_1$ acts to the left and $\vec{F}_3$ acts downward, so no further vector reductions are necessary. However, $\vec{F}_2$ acts at some angle toward the lower left, given by θ. To find the angle, we use the sides of the given rectangle to determine that $\tan \theta = 5/7$, which implies that $\theta = 35.53°$. Both components of $\vec{F}_2$ will be negative by our sign convention of Chapter 1.

We can now write:

$$F_{2x} = -10.95 \cos 35.53 = -8.91 \text{ N}$$

and

$$F_{2y} = -10.95 \sin 35.53 = -6.36 \text{ N}$$

The resultant or net force acting on the lower-left-corner charge is the vector sum of these charges. We can find the components of this force by simply adding up the forces in the same directions:

$$F_x = -16.53 - 8.91 = -25.44 \text{ N}$$

$$F_y = -32.40 - 6.36 = -38.76 \text{ N}$$

Here is a summary of all the forces and their components in a graphic organizer, a highly recommended template for students to use during problem-solving:

Total Force	X component	Y component
F_1 = 16.53 N	−16.53 N	0 N
F_2 = 10.95 N	−10.95 cos(35.53) N	−10.95 sin(35.53) N
F_3 = 32.4 N	0 N	−32.4 N
Total	−25.44 N	−38.76 N

The magnitude of the net force F is given by the Pythagorean theorem and is equal to 46.3 N. The direction of this force is given by

$$\tan\phi = \frac{F_y}{F_x} = \frac{38.76}{25.44} = 1.5$$

This implies that $\phi = 56.7°$ relative to the negative x-axis.

2. The only conclusion you can make is that the sphere is either "neutral" or oppositely charged compared to the rod.

3. If your car is hit by lightning, the charges are distributed around the outside and then dissipated away. In a sense, this acts like an electrostatic shield.

4. The formula we want to use is

$$R = \frac{\rho L}{A}$$

From Table 9.1, the resistivity of platinum at 20°C is equal to

$$\rho = 10.4 \times 10^{-8} \Omega \cdot \text{m}$$

The resistance needed is $R = 60\ \Omega$. We need the cross-sectional area. The diameter of the wire is 0.10 mm, so the radius of the wire must be equal to

$$r = \frac{\text{diameter}}{2} = 0.05 \text{ mm} = 0.00005 \text{ m}$$

The cross-sectional area is

$$A = \pi r^2 = \pi(0.00005 \text{ m})^2 = 7.85 \times 10^{-9} \text{ m}^2$$

The length of the wire is therefore

$$60\ \Omega = \frac{(10.4 \times 10^{-8}\ \Omega \cdot \text{m})L}{(7.85 \times 10^{-9}\text{ m}^2)}$$

$$L = 4.53\text{m}$$

5. (a) The first thing we need to do is to reduce the series portion of the parallel branch. The equivalent resistance there is $8\,\Omega + 2\,\Omega = 10\,\Omega$. Now, this resistance is in parallel with the other 10-Ω resistor. The equivalent resistance is found to be $5\,\Omega$ for the entire parallel branch. This equivalent resistance is in series with the other 5-Ω resistor, making a total equivalent resistance of $10\,\Omega$.

    (b) We can determine the circuit current using Ohm's law:

    $$I = \frac{V}{R} = \frac{40}{10} = 4\,\text{A}$$

    (c) To determine the reading of ammeter A, we must first understand that the potential drop across the 5-Ω resistor on the bottom is equal to 20 V since

    $$V = IR = (4)(5) = 20\,\text{V}$$

    Now, the equivalent resistance of the parallel branch is also $5\,\Omega$, and therefore across the entire branch there is also a potential difference of 20 V (in a series circuit, the voltages must add up—in this case, to 40 V). In a parallel circuit, the potential difference is the same across each portion. Thus, the 10-Ω resistor has 20 V across it. Using Ohm's law, we find that this implies a current reading of 2 A for ammeter A.

    (d) If, from part (c), ammeter A reads 2 A, the top branch must be getting 2 A of current as well (since the source current is 4 A). Thus, using Ohm's law, we can find the voltage across the 8-Ω resistor:

    $$V = IR = (2)(8) = 16\ \text{V}$$

6. We start with the regular formula for two resistors in parallel and solve:

    $$\frac{1}{R_{eq}} = \frac{1}{R_1} + \frac{1}{R_2}$$

    $$\frac{1}{R_{eq}} = \frac{R_2}{R_2 R_1} + \frac{R_1}{R_1 R_2}$$

    $$\frac{1}{R_{eq}} = \frac{R_1 + R_2}{R_1 R_2}$$

    $$R_{eq} = \frac{R_1 R_2}{R_1 + R_2}$$

7. When two batteries are connected in parallel, the emf of the system is not changed. The storage capacity of the battery system, however, is increased since the parallel connection of the two batteries is similar to the parallel connection of two capacitors.

8. An equivalent resistance of $1\,\Omega$ can be achieved by connecting two 2-Ω resistors in parallel. An equivalent resistance of $3\,\Omega$ can be achieved by connecting two 2-Ω resistors in parallel and then connecting this system to one 2-Ω resistor in series. The addition of one more 2-Ω resistor in series to the combination above achieves an equivalent resistance of $5\,\Omega$.

9. Lightbulbs produce their luminous energy by heating up the filament inside them. Thus, after a while, this increase in temperature reduces the current flowing through them since the resistance has increased.

Waves and Sound

10

KEY CONCEPTS

→ **PULSES**

→ **WAVE MOTION**

→ **TYPES OF WAVES**

→ **STANDING WAVES AND RESONANCE**

→ **SOUND**

→ **THE DOPPLER EFFECT**

PULSES

A pulse is a single vibratory disturbance in a medium. An example of a pulse is seen in Figure 10.1. If a string fixed at both ends and made taut, under a tension T (in newtons), is given a quick up and down snap, an upwardly pointing pulse is directed from the left to the right. The pulse appears to travel with a velocity $\vec{v}$ down the string. In actuality, the energy transferred to the string causes segments to vibrate up and down successively. This effect produces the illusion of a continuous pulse. Only energy is transferred by the pulse, and because the vibration is perpendicular to the apparent direction of motion, the pulse is said to be **transverse**. The **amplitude** of the pulse is the displacement of the disturbance above the level of the string.

Transverse pulse

Figure 10.1

If the tension in the string is increased, the velocity of the pulse increases. If a heavier string is used (greater mass per unit length), the pulse appears to move slower. Careful measurements of these observations lead to the following equation for the velocity of the pulse:

$$v = \sqrt{\frac{T}{M / \ell}}$$

where T is the tension in newtons, M is the mass in kilograms, and ℓ is the length of the string in meters. In general, the velocity of pulses and waves is dependent on the **medium**. The medium is the material through which the waves and pulses are traveling.

When the pulse reaches one boundary, the energy transferred is directed upward against the wall, creating an upward force. Since the wall is rigid, the reaction to this action is a downward force. The string responds by being displaced downward and is reflected back. The result of this boundary interaction, illustrated in Figure 10.2, is observed as an inversion of the pulse upon reflection. Some energy is lost in the interaction, but most is reflected back with the pulse. This is known as "phase inversion."

TIP

Waves transfer only energy. The medium returns to its original state after the vibrations pass through.

(a) Incidence

(b) Reflection

Figure 10.2

At a nonrigid boundary, say between two strings of different masses (but equal tensions), a different effect occurs. When the pulse reaches the nonrigid boundary, the upward displacement causes the second string to also be displaced upward. The magnitude of the second displacement depends on the mass density and tension in the second string. Thus, the transmitted pulse has an upward orientation and may travel with a larger or smaller velocity, depending, in this case, on the mass density.

The reflected pulse will be upward in orientation since no inversion takes place with a nonrigid boundary unless the difference between the two media (as measured by the second string's inertia) is large enough to behave as a semirigid boundary. As an example, consider a light string attached to a heavy rope, with the pulse traveling from the string to the rope. Figure 10.3 illustrates both cases.

Incident pulse

Reflected pulse
Transmitted pulse

Case I: Rigid reflection

Incident pulse

Reflected pulse
Transmitted pulse

Case II: Nonrigid reflection

Figure 10.3

Since pulses transmit only energy, when two pulses interact, they obey a different set of physical laws than when two pieces of matter interact. Pulses are governed by the **principle of superposition**, which states that, when two pulses interact at the same point and at the same time, the interaction produces a single pulse whose amplitude displacement is equal to the sum of the displacements of the original two pulses. After the interaction, however, both pulses continue in their original directions of motion, unaffected by the interaction. Sometimes, this interaction is called **interference**.

When a pulse with an upward displacement interacts with another pulse with an upward displacement, the resulting pulse has a displacement larger than that of either original pulse (equal to the numerical sum of these pulses). This interaction is called constructive interference.

When a pulse with an upward displacement interacts with a pulse that has a downward displacement, the interaction is called **destructive interference**, and the resulting pulse has a smaller displacement than either original pulse (equal to the numerical difference of these pulses). Figure 10.4 illustrates these two types of interferences; a and b designate the two original amplitudes, and p is a point on the string.

Before superposition	Before superposition
During superposition	During superposition
After superposition	After superposition
(a) Constructive interference	(b) Destructive interference

Figure 10.4

In the destructive interference case, it is possible, if the amplitudes are equal, that the interaction will momentarily cancel out both pulses. In any case, the orientation of the resulting pulse depends on the magnitude of each amplitude.

WAVE MOTION

If a continuous up and down vibration is given to the string in Figure 10.1, the resulting set of transverse pulses is called a **wave train** or just a **series of waves**. Whereas the pulse was just an upward- or downward-oriented, transverse disturbance, a wave consists of a complete up and down segment. Waves are thus periodic disturbances in a medium. The number of waves per second is called the **frequency**, designated by the letter f and expressed in units of reciprocal seconds (s^{-1}) or **hertz** (Hz). The time required to complete one wave cycle is called the **period**, designated by the capital letter T and expressed in seconds. The frequency and period of a wave are reciprocals of each other, so that $f = 1/T$.

Like pulses, waves transmit only energy. The particles in the medium vibrate only up and down as the transverse waves approach and pass a given point. Since there is a continuous series of waves, many points in the medium will be moving up or down in unison. The **phase** of a wave is defined to be the relative position of a point on a wave with respect to another point on the same wave.

The distance between any two successive points in phase is a measure of the **wavelength**, designated by the Greek letter λ (lambda) and measured in meters. The peaks of the wave are called **crests**; the valleys, **troughs**. Thus, one wavelength can also be measured as the distance between any two successive crests or troughs. The **amplitude**, A, of the wave is the distance, measured in meters, of the maximum displacement, up or down, above the normal rest position (equilibrium level). All waves share the following relationship between wavelength, frequency, and the speed of the wave: $v = f\lambda$.

Figure 10.5 illustrates a typical transverse wave.

Figure 10.5 Transverse Wave

The sinusoidal nature of the wave is similar to the graph of displacement versus time for a mass on a spring undergoing simple harmonic motion. We recall from Chapter 8 that the maximum energy in a simple harmonic motion system is directly proportional to the square of the amplitude. This is indeed the case with a simple wave motion like the one described in Figure 10.5. In this illustration we can identify points A and A' as all in phase. Points B and B' are also in phase, and the distance from B to B' can also be used to measure the wavelength.

Since each wave carries energy, it should not be surprising that the frequency is also proportional to the energy of a wave. The amplitude is related to what we might consider the wave's **intensity**. In sound, this property might be called **volume**. The frequency, however, measures the **pitch** of the wave. Waves of higher frequency transmit, at the same amplitude, more energy per second than waves of lower frequency.

There are many different kinds of frequencies that interact with human senses. It is worthwhile, therefore, to recall that the prefix *kilo-* represents 10^3, the prefix *mega-* represents 10^6, and the prefix *giga-* represents 10^9.

TYPES OF WAVES

In physics we usually deal with two types of waves, mechanical and electromagnetic. Waves that result from the vibration of a physical medium (a drum, a string, water, etc.) are called **mechanical waves**. Waves that result from electromagnetic interactions (light, X-rays, radio waves, etc.) are called **electromagnetic waves**. These electromagnetic waves are special since they do not require a physical medium to carry them. Also, all electromagnetic waves are transverse in nature.

Another way to set up periodic disturbances in the spring is to pull the spring back and forth along its longitudinal axis. The resulting disturbances consist of regions of compressions and expansions that appear to travel parallel to the disturbances themselves. These waves are called **longitudinal** (Figure 10.6) or **compressional** waves. Sound is an example of a longitudinal wave. The vibrations of an object in the air create pressure differences that alternately expand and contract the air and that, when impacted on our ears, create the phenomenon known as **sound**.

> ### REMINDER
> Make sure you know the differences between **transverse** and **longitudinal** waves. You should also know the differences between **mechanical** and **electromagnetic** waves.

Figure 10.6 Longitudinal Wave

STANDING WAVES AND RESONANCE

Consider Figure 10.1 again. There are several ways in which we can make the taut string vibrate transversely. One way is to make the entire string move up and down as a single unit. This is the easiest frequency mode of vibration possible and is called the **fundamental mode**. If we increase the frequency, something interesting happens. The waves reaching a boundary reflect off it inverted and match perfectly, in amplitude and frequency, the remaining incoming waves. In this situation, the apparent horizontal motion of the wave stops and we have a **standing wave**, which appears to be segmented by **nodal points**. These nodal points correspond to points where no appreciable displacement takes place.

Figure 10.7 shows a standing wave in various vibrational modes. Figure 10.7a illustrates the fundamental mode. In Figure 10.7b we see the second mode, in which one nodal point appears. This nodal point occurs at the one-half wavelength mark, so that, for a string of length ℓ, $\lambda = \ell$ (at this frequency). In the fundamental mode of Figure 10.7a, $\lambda = 2\ell$. In Figure 10.7c, there are two nodal points and $\lambda = 2\ell/3$.

This standing-wave pattern is a form of interference, and with sound the nodal points would be determined by a significant drop in intensity. With light, we would observe a darker region relative to the surrounding area. With a string, we would see the characteristic segmented pattern shown in Figure 10.7. Note that each segment is half a wavelength, bounded by 2 nodes with an **antinode** in the middle.

a $\ell = \dfrac{\lambda}{2}$

b $\ell = \lambda$

c $\ell = \dfrac{3}{2}\lambda$

d $\ell = 2\lambda$

Figure 10.7

TIP

Nodes are destructive interference.

Antinodes are constructive interference.

Standing waves can be manipulated so that the maximum points continuously reinforce themselves. This buildup of wave energy due to the constructive interference of standing waves is called **resonance**. Resonance can also be induced by an external agent.

All objects have a natural vibrating frequency. When a glass or bell is struck, only one fundamental characteristic frequency of vibration is heard (actually, the sound is a complex mixture of vibrations). If, however, two identical tuning forks, A and B, are held close together, then the phenomenon of resonance can be observed. If tuning fork A is struck, the waves emanating from it strike fork B. Since the frequency of these waves match the natural vibrating frequency of fork B, that fork will begin to vibrate (see Figure 10.0).

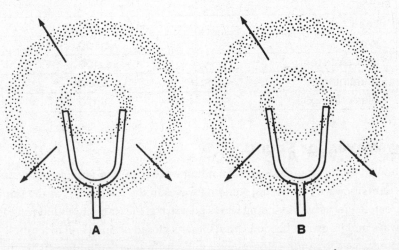

A   B

Figure 10.8 Resonance in Two Identical Tuning Forks

Resonance also explains how an opera singer can shatter a crystal wine glass. More important, it explains why soldiers are ordered to "break march" when crossing a bridge. The rhythmic marching can set up a resonance vibration and perhaps collapse the bridge. In November 1940, a resonance vibration caused by a light gale wind generated catastrophic torsional vibrations in the Tacoma Narrows Bridge in Washington State. The violent convulsions were photographed and are shown as a classic illustration of the principle of resonance. Nicknamed "Galloping Gertie," the bridge has become a testament to the need for engineers to be very careful when considering the possible effects of resonance.

SOUND

Sound is a longitudinal mechanical wave. Air molecules are alternately compressed and spaced out (rarefied) as pressure differences move through the air. When lightning occurs, the expansion of air due to the very high temperature of the lightning bolt creates the violent sound known as thunder.

Since sound is "carried" by air molecules, it is temperature dependent. At 0 degree Celsius, the velocity of sound in air is 331 meters per second. For each 1-degree rise in temperature, the velocity of sound increases by 0.6 meter per second. The study of sound is called **acoustics**, and longitudinal waves in matter are sometimes termed **acoustical waves**.

The ability of sound waves to pass through matter depends on the structure of the matter involved. Thus, sound travels slower in a gas, in which the molecules have more random motion, than in a liquid or a solid, which has a more "rigid" structure. In Table 10.1 the velocities of sound in selected substances are listed.

Table 10.1. Speed of Sound in Selected Substances

Substance	Speed (m/s)
Gases (0°C)	
Carbon dioxide	267
Air	343
Helium	1,007
Liquids (25°C)	
Ethyl alcohol	1,207
Water, pure	1,498
Water, sea	1,531
Solids	
Lead	1,200
Wood	~4,300
Iron and steel	~5,000
Aluminum	5,100
Glass (Pyrex)	5,170

Standing Waves for Sound

Just as we showed for mechanical waves, standing waves can also form sound waves. In fact, all musical instruments work by creating standing waves in the instrument at the frequency of the note being created. Just as with standing waves on a string, the standing sound wave is an interference of the wave and its own reflection (from either a closed or open end of a tube). The progression of standing waves for a tube closed on both ends (or open on both ends) is the same as for those of the string in Figure 10.7, but the tube that is open on only one end has a slightly different progression:

Lowest harmonic: $\ell = 1/4\lambda$

Next harmonic: $\ell = 3/4\lambda$

Next harmonic: $\ell = 5/4\lambda$

Etc.

Open-Closed Pipe
FIRST HARMONIC

Open-Closed Pipe
THIRD HARMONIC

Why do some sources picture the open ends of pipes as nodes and others as antinodes? It depends on whether they are modeling the standing sound wave as a pressure wave or as a displacement of molecules wave. Closed ends are nodes for displacement but antinodes for pressure. Open ends are antinodes for displacement but nodes for pressure.

SAMPLE PROBLEM

Consider a student blowing over the top of an empty bottle and hearing a specific pitch at the frequency of the standing air wave produced by the column of air in the bottle.

Pictured above is a variation of this problem where the frequency is set but the column of air is adjusted. What is the wavelength and frequency of this dominant sound wave in terms of the height, h, of the empty bottle? (Hint: The loudest standing wave in the air column in the bottle will experience a node at one end and an antinode at the other.)

Solution

If a standing wave has a node at one end and an antinode at the other, it will look like exactly half of the first standing wave in Figure 10.7. This means that the height of the bottle is one-fourth of a wavelength:

$$\text{wavelength} = 4h$$

To find the pitch (or frequency) of the sound wave, use the general relationship:

$$\text{wavelength} \times \text{frequency} = \text{wave speed}$$

In this case, use the speed of sound in air (approximately 340 m/s). Therefore, assuming h is in meters, the frequency of this fundamental pitch can be calculated:

$$f = \frac{340}{4h}$$

Note that many musical instruments are constructed to be quarter-wavelength pipes that are closed on only one end, just like in this example.

TIP

Wave *speed* is set by the medium.

Wave *frequency* is set by the source.

Human hearing can detect sound in a range from 20 to 20,000 hertz. Sound waves exceeding a frequency of 20,000 hertz are called **ultrasonic** waves. The amplitude of the sound wave is the intensity of **loudness** of the wave, while the frequency relates to the **pitch**. When two sound waves interfere, the regions of constructive and destructive interference produce the phenomena known as **beats**. The number of beats per second is equal to the frequency difference.

Interference, for any kind of wave, is dependent on what is called the **path-length difference**. Consider two point sources of sound, A and B, and a receiver some distance away. If the distance from source A to the receiver is ℓ_A and the distance from source B to the receiver is ℓ_B, the receiver will be at a point of constructive interference if the path-length difference, $\ell_A - \ell_B$, is equal to a whole multiple of the wavelength, $n\lambda$. Destructive interference will occur if the path-length difference is equal to an odd multiple of the half-wavelengths, $(n + 1/2)\lambda$.

Another phenomenon associated with interference is **diffraction**. When a wave encounters a boundary, the wave appears to bend around the corners of the boundary. This effect, known as **diffraction**, occurs because at the corners, the wave behaves like a point source and creates circular waves. These waves, because of their shape, reach behind the corners, and this continuous effect gives the illusion of wave bending. In Figure 10.9a, we observe a series of straight waves, made in a water tank, and the resulting diffraction at a corner. Figure 10.9b shows the diffraction of waves as they pass through a narrow opening. "Narrow" is relative to the wavelength of wave present. Waves do not show significant diffraction through openings much larger than their wavelength.

(a) Diffraction around a corner   (b) Diffraction through a narrow opening

Figure 10.9

THE DOPPLER EFFECT

Almost everyone has had the experience of hearing a siren pass by. Even though the siren is emitting a sound at a single frequency, the changing position of the siren, relative to the hearer, produces an apparent change in the frequency. As the siren approaches, the pitch is increased; as the siren passes, the pitch is decreased. This phenomenon is called the **Doppler effect**.

> Resonance, interference, diffraction, and the Doppler shift are phenomena associated with **ALL** waves!

In Figure 10.10, the source is moving, let's say, toward point A. Then, for each period of time T between waves, the circular waves will not be concentric. In other words, the spacing between each two successive waves will be reduced by an amount equal to the distance traveled by the source in time T. Thus, an apparent increase in frequency is experienced at point A, while at point B there is an apparent decrease in frequency.

Figure 10.10

This Doppler shift, as it is known, is the result of the relative motion between the source of the wave and the receiver or observer. This can be accomplished by any combination of source and observer movement. As long as the two are approaching, the received frequency will be higher.

When the two are getting farther apart, the received frequency will be lower. The amount of the shift in frequency is related to the relative velocity of the two compared with the velocity of the wave itself.

SUMMARY

- Mechanical waves are periodic vibratory disturbances in a medium.
- A pulse is a single vibratory disturbance.
- A pulse or wave is transverse if its vibrations are perpendicular to the direction of propagation.
- A pulse or wave is longitudinal if its vibrations are parallel to the direction of motion.
- Waves transmit only energy. The principle of superposition states that waves can be added together.
- The amplitude of a wave is its maximum displacement from its equilibrium position.
- The frequency of a wave is the number of cycles per second.
- The period of a wave is the time that it takes to complete one cycle and is equal to the reciprocal of the frequency.
- Particles in a medium that display the same state of motion, simultaneously, are said to be in phase.
- The "wavelength" of a wave is the distance between any two successive points in phase.
- The wave velocity is equal to the product of the frequency times the wavelength.
- A standing wave is a wave in which incident and reflected waves combine to produce a wave that appears to be "standing" in one place.
- Standing wave nodes are points where there is no appreciable displacement. Nodes occur every half-wavelength.
- Resonance is the maximum transfer of energy from one body to another that shares the same natural vibrating frequency.
- Reflection is the returning of a wave in the opposite direction due to a difference in media at a boundary.
- Refraction is the bending of a wave as it enters a new medium, at an oblique angle, with different propagation speed.
- Diffraction is the bending of waves around obstacles in a given medium or the outward speed as the wave passes through small openings.
- Interference is the superposition of two waves of identical or opposite phase to produce constructive or destructive interference, respectively.
- Sound is a longitudinal mechanical wave.
- The Doppler effect states that when a source is moving relatively toward an observer, there is an apparent increase in frequency. If the relative motion is away from source or observer, there is an apparent decrease in frequency.

Problem-Solving Strategies for Waves and Sound

Solving wave motion problems involves remembering the basic physical concepts involved. For transverse waves, the only particle motion in the medium will be up and down. If a series of waves is presented, you must resist the temptation of viewing labeled points as beads on a wire. As a wave approaches a point, it will first go up and then go down. Points that are in phase are either going up at the same time or going down at the same time.

Sound waves travel through a medium, and their speed is determined by the molecular structure of the matter involved. In air, the speed of sound increases with increasing temperature. Sound wave interference is manifested by the beat pulsations heard when two frequencies are experienced simultaneously.

PRACTICE EXERCISES

Multiple-Choice

1. A stretched string has a length of 1.5 m and a mass of 0.25 kg. What must be the tension in the string in order for pulses in the string to have a velocity of 5 m/s?

   (A) 2.45 N
   (B) 12.5 N
   (C) 4.2 N
   (D) 150 N

2. A stretched string is vibrated in such a way that a standing wave with two nodes appears. The distance between the nodes is 0.2 m. What is the wavelength of the standing wave?

   (A) 0.4 m
   (B) 0.3 m
   (C) 0.2 m
   (D) 0.6 m

3. What is the wavelength of sound produced at a frequency of 300 Hz when the air temperature is 20°C?

   (A) 1.10 m
   (B) 1.30 m
   (C) 0.80 m
   (D) 1.14 m

4. Two tuning forks are vibrating simultaneously. One fork has a frequency of 256 Hz; the other, a frequency of 280 Hz. The number of pulsational beats per second heard will be

   (A) 256
   (B) 536
   (C) 1.1
   (D) 24

5. In a stretched string with a constant tension T, as the frequency of the waves increases, the wavelength

   (A) increases
   (B) decreases
   (C) remains the same
   (D) increases, then decreases

6. What is the period of a wave that has a frequency of 12,000 Hz?

   (A) 36.25 s
   (B) 0.000083 s
   (C) 0.0275 s
   (D) 12,000 s

7. Sound waves travel fastest in

   (A) a vacuum
   (B) air
   (C) water
   (D) wood

8. The amplitude of a sound wave is related to its

   (A) pitch
   (B) loudness
   (C) frequency
   (D) resonance

9. At 25°C, a sound wave takes 3 s to reach a receiver. How far away is the receiver from the source?

   (A) 1,038 m
   (B) 993 m
   (C) 1,215 m
   (D) 1,068 m

Free-Response

1. What must be the tension in a 0.25-m string with a mass of 0.30 kg so that its fundamental frequency mode is 400 Hz?

2. Why should people not march in unison when crossing over a bridge?

3. Apartment dwellers are used to hearing the "boom-boom" sound of bass tones when neighbors play their stereos too loudly. What can account for this phenomenon?

4. People who live near airports claim that they hear planes that are on the ground more often during the summer than at other times of the year. Explain why this might occur.

ANSWERS EXPLAINED

Multiple-Choice Problems

1. **(C)** The formula for wave velocity in a string is

$$v = \sqrt{\frac{T}{M/\ell}}$$

Using the given information and solving for tension leads to $T = 4.2$ N.

2. **(A)** Standing-wave nodes occur at half-wavelength intervals. Thus,

$$\lambda = 2(0.2) = 0.4 \text{ m}$$

3. **(D)** At 20°C, the speed of sound is $330 + (0.6)(20) = 343$ m/s. Since $v = f\lambda$, we get a wavelength of $\lambda = 1.14$ m.

4. **(D)** The number of beats per second is equal to the frequency difference: $280 - 256 = 24$.

5. **(B)** Since $v = f\lambda$ and since, for constant tension, the velocity is constant, as the frequency increases, the wavelength decreases.

6. **(B)** Frequency and period are reciprocals of each other. Thus

$$T = \frac{1}{12,000} = 0.000083 \text{ s}$$

7. **(D)** Because of molecular arrangement, sound travels faster in solids than in liquids or gases.

8. **(B)** Amplitude is related to the intensity of a wave. In sound, wave intensity is perceived as loudness.

9. **(A)** At 25°C, the speed of sound is $331 + (0.6)(25) = 346$ m/s. Since $x = vt$,

$$x = (346)(3) = 1,038 \text{ m}$$

Free-Response Problems

1. At the first harmonic, or fundamental frequency, we have

$$\lambda = 2\ell = (2)(0.25) = 0.5 \text{ m}$$

Now, $v = f\lambda$, which means that

$$v = (400)(0.5) = 200 \text{ m/s}$$

We can now use the formula for the velocity in a fixed string:
$$v = \sqrt{\frac{T}{M/\ell}}$$
Solving for the tension gives $T = 48,000$ N.

2. People should not march in unison when crossing over a bridge because the uniform vibrations may induce resonance, which can lead to destructively large amplitudes.

3. The resonant frequency of apartment building walls is quite low, and so low-frequency bass tones resonate through the walls as the "boom-boom" sound so familiar to apartment building dwellers.

4. Since the velocity of sound increases with temperature, the hot summer air carries the sound waves farther.

Practice Tests

ANSWER SHEET
Practice Test 1

1. Ⓐ Ⓑ Ⓒ Ⓓ
2. Ⓐ Ⓑ Ⓒ Ⓓ
3. Ⓐ Ⓑ Ⓒ Ⓓ
4. Ⓐ Ⓑ Ⓒ Ⓓ
5. Ⓐ Ⓑ Ⓒ Ⓓ
6. Ⓐ Ⓑ Ⓒ Ⓓ
7. Ⓐ Ⓑ Ⓒ Ⓓ
8. Ⓐ Ⓑ Ⓒ Ⓓ
9. Ⓐ Ⓑ Ⓒ Ⓓ
10. Ⓐ Ⓑ Ⓒ Ⓓ
11. Ⓐ Ⓑ Ⓒ Ⓓ
12. Ⓐ Ⓑ Ⓒ Ⓓ
13. Ⓐ Ⓑ Ⓒ Ⓓ

14. Ⓐ Ⓑ Ⓒ Ⓓ
15. Ⓐ Ⓑ Ⓒ Ⓓ
16. Ⓐ Ⓑ Ⓒ Ⓓ
17. Ⓐ Ⓑ Ⓒ Ⓓ
18. Ⓐ Ⓑ Ⓒ Ⓓ
19. Ⓐ Ⓑ Ⓒ Ⓓ
20. Ⓐ Ⓑ Ⓒ Ⓓ
21. Ⓐ Ⓑ Ⓒ Ⓓ
22. Ⓐ Ⓑ Ⓒ Ⓓ
23. Ⓐ Ⓑ Ⓒ Ⓓ
24. Ⓐ Ⓑ Ⓒ Ⓓ
25. Ⓐ Ⓑ Ⓒ Ⓓ
26. Ⓐ Ⓑ Ⓒ Ⓓ

27. Ⓐ Ⓑ Ⓒ Ⓓ
28. Ⓐ Ⓑ Ⓒ Ⓓ
29. Ⓐ Ⓑ Ⓒ Ⓓ
30. Ⓐ Ⓑ Ⓒ Ⓓ
31. Ⓐ Ⓑ Ⓒ Ⓓ
32. Ⓐ Ⓑ Ⓒ Ⓓ
33. Ⓐ Ⓑ Ⓒ Ⓓ
34. Ⓐ Ⓑ Ⓒ Ⓓ
35. Ⓐ Ⓑ Ⓒ Ⓓ
36. Ⓐ Ⓑ Ⓒ Ⓓ
37. Ⓐ Ⓑ Ⓒ Ⓓ
38. Ⓐ Ⓑ Ⓒ Ⓓ
39. Ⓐ Ⓑ Ⓒ Ⓓ

40. Ⓐ Ⓑ Ⓒ Ⓓ
41. Ⓐ Ⓑ Ⓒ Ⓓ
42. Ⓐ Ⓑ Ⓒ Ⓓ
43. Ⓐ Ⓑ Ⓒ Ⓓ
44. Ⓐ Ⓑ Ⓒ Ⓓ
45. Ⓐ Ⓑ Ⓒ Ⓓ
46. Ⓐ Ⓑ Ⓒ Ⓓ
47. Ⓐ Ⓑ Ⓒ Ⓓ
48. Ⓐ Ⓑ Ⓒ Ⓓ
49. Ⓐ Ⓑ Ⓒ Ⓓ
50. Ⓐ Ⓑ Ⓒ Ⓓ

Practice Test 1

SECTION I: MULTIPLE-CHOICE

Time: 90 minutes
50 questions

> **DIRECTIONS:** Each of the questions or incomplete statements below is followed by four suggested answers or completions. Select the one (or two where indicated) that is best in each case. You have 90 minutes to complete this portion of the test. You may use a calculator and the information sheets provided in the appendix.

1. Two objects are thrown vertically upward from the same initial height. One object has twice the initial velocity of the other. Neglecting any air resistance, the object with the greater initial velocity will rise to a maximum height that is

   (A) twice that of the other object, assuming they have the same mass
   (B) twice that of the other object, independent of their masses
   (C) four times that of the other object, assuming they have the same mass
   (D) four times that of the other object, independent of their masses

2. A 2-kilogram cart has a velocity of 4 meters per second to the right. It collides with a 5-kilogram cart moving to the left at 1 meter per second. After the collision, the two carts stick together. Can the magnitude and the direction of the velocity of the two carts after the collision be determined from the given information?

   (A) No, since the collision is inelastic, we must know the energy lost.
   (B) Yes, the collision is elastic: 3/7 m/s left.
   (C) Yes, the collision is inelastic: 3/7 m/s right.
   (D) Yes, the speed is not 3/7 m/s.

3. Projectile *X* is launched at a 30-degree angle above the horizon with a speed of 100 m/s. Projectile *Y* is launched at a 60-degree angle with the same speed. Which of the following correctly compares the horizontal range and maximum altitude obtained by these two projectiles?

	Range	Altitude
(A)	*X* goes farther	*X* goes higher
(B)	*Y* goes farther	*Y* goes higher
(C)	*X* goes farther	*Y* goes higher
(D)	*X* and *Y* equal	*Y* goes higher

4. A 5-kg mass is sitting at rest on a horizontal surface. A horizontal force of 10 N will start the mass moving. What is the best statement about the coefficient and type of friction between the mass and the surface?

(E) >0.20 static
(F) <0.20 static
(G) <0.20 kinetic
(H) >0.20 kinetic

5. Which of the following graphs represents an object *moving* with no net force acting on it?

(A)

(B)

(C)

(D)

6. A projectile is launched horizontally with an initial velocity v_0 from a height h. If it is assumed that there is no air resistance, which of the following expressions represents the vertical position of the projectile? In other words, $y(x) = ?$

(A) $h - gv_0^2/2x^2$
(B) $h - gv/2v_0^2$
(C) $h - gx^2/2v_0^2$
(D) $h - gx^2/v_0^2$

QUESTIONS 7 AND 8 ARE BASED ON THE INFORMATION AND DIAGRAM BELOW:

A 0.4-kilogram mass is oscillating on a spring that has a force constant of $k = 1,000$ newtons per meter.

7. Which of the following measurements would allow you to determine the maximum velocity experienced by the mass?

(A) No additional information is required.
(B) Minimum velocity
(C) Maximum acceleration
(D) None of these would allow you to determine maximum velocity.

8. Which of the following statements concerning the oscillatory motion described above is correct? (All statements refer to magnitudes.)

(A) The maximum velocity and maximum acceleration occur at the same time.
(B) The maximum velocity occurs when the acceleration is a minimum.
(C) The velocity is always directly proportional to the displacement.
(D) The maximum velocity occurs when the displacement is a maximum.

9. A bullet of known mass (m_1) is fired vertically into an initially stationary wood block of known mass (m_2). The resulting wood + bullet combined system is then measured to rise to a maximum height of h. Can the initial speed of the bullet be calculated from this information?

(A) Yes. Solve $(m_1 + m_2)gh = \frac{1}{2}(m_1)v^2$.
(B) Yes. Solve the momentum conservation of collision first and the energy conservation of the rising combination second.
(C) No. We don't know if momentum is conserved during this collision.
(D) No. We don't know enough details about the mechanical energy lost during the collision.

10. A 10-kg mass is being pushed horizontally by a constant force along a rough surface ($\mu_k = 0.1$) at constant velocity. Which of the following is the best statement regarding the constant force (f)?

(A) $f = 10$ N
(B) $f > 10$ N
(C) $f < 10$ N
(D) $f > 1$ N

11. A 10-newton force is applied to two masses, 4 kilograms and 1 kilogram, respectively, that are in contact as shown below. The horizontal motion is along a frictionless plane. What is the magnitude of the contact force between the two masses?

(A) 10 N

(B) 8 N

(C) 6 N

(D) 2 N

12. An object with mass m is dropped from height h above the ground. While neglecting air resistance, which formula best describes the power generated if the object takes time t to fall?

(A) mgh

(B) $mght$

(C) $mg^2t/2$

(D) mgh/t

13. A 1,500-kilogram car has a velocity of 25 meters per second. If it is brought to a stop by a nonconstant force in 10 to 15 seconds, can the magnitude of the impulse applied be determined?

(A) Yes, it is 37,500 N · s.

(B) No, you need to know the details about the nonconstant force.

(C) No, you must know the exact duration of the impulse.

(D) No, you must know the average force during and the duration of the impulse.

14. A block of mass M rests on a rough incline, as shown below. The angle of elevation of the incline is increased until an angle of θ is reached. At that angle, the mass begins to slide down the incline. Which of the following is an expression for the coefficient of static friction μ?

(A) $\tan \theta$

(B) $\sin \theta$

(C) $\cos \theta$

(D) $1/(\cos \theta)$

15. A pendulum of a given length swings back and forth a certain number of times per second. If the pendulum now swings back and forth the same number of times but in twice the time, the length of the pendulum should be

   (A) doubled
   (B) quartered
   (C) quadrupled
   (D) halved

16. This graph of force versus time shows how the force acts on an object of mass m for a total time of T seconds. If the mass begins at rest, which is the correct method to find the final speed of the mass?

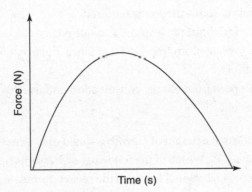

   (A) Average value of this graph times total time divided by mass
   (B) Area under this graph divided by mass
   (C) Since the final force is zero, the object is at rest after time T
   (D) Average slope of this graph divided by mass

17. A child of unknown mass is on a swing of unknown length that varies in height from 75 cm at its lowest height above the ground to a maximum height of 225 cm above the ground. Is there enough information to find the speed of the swing at its lowest point?

   (A) No, the child's mass must be known.
   (B) No, the length of the swing must be known to determine the centripetal acceleration.
   (C) Yes, it is 5.5 m/s.
   (D) Yes, it is 4 m/s.

18. The gravitational force of attraction between two identical masses is 36 N when the masses are separated by a distance of 3 m. If the distance between them is reduced to 1 m, which of the following is true about the net gravitational field strength due to both masses being at the halfway point?

   (A) It is 9 times stronger total.
   (B) Not enough information is given to determine net gravitational field strength.
   (C) Each mass's gravitation is 9 times stronger, so the net gravitational field strength is 18 times stronger.
   (D) It is zero.

19. A 200-kilogram cart rests on top of a frictionless hill as shown below. Can the impulse required to stop the cart when it is at the top of the 10-meter hill be calculated?

(A) No, more information about friction is required.

(B) No, more information about the impulse is required.

(C) Yes, calculate the velocity from free-fall kinematics and use this velocity in the change in momentum equation.

(D) Yes, calculate the speed from energy conservation and use this speed in the change in momentum equation.

20. An object of mass m starts at a height of H_1 with a speed of v_1. A few minutes later, it is at a height of H_2 and a speed v_2. Which of the following expressions best represents the work done to the mass by nongravitational forces to the object during this time?

(A) $mg(H_2 - H_1) + \frac{1}{2}m(v_2^2 - v_1^2)$

(B) $mg(H_2 - H_1) - \frac{1}{2}m(v_2^2 - v_1^2)$

(C) $\frac{1}{2}m(v_2^2 - v_1^2)$

(D) $\frac{1}{2}m(v_1^2 - v_2^2)$

21. An object with 0.2-kg mass is pushed down vertically onto an elastic spring ($k = 20$ N/m). The spring is compressed by 20 cm and then released such that the object will fly off. Which of the following will have the largest effect on increasing the maximum height the object will fly? (Assume no air resistance.)

(A) Halving the mass

(B) Doubling the compression distance

(C) Using a spring with a spring constant twice as big

(D) Doing the same experiment on a different planet with half the gravitational field strength

22. A cart with a mass of m needs to complete a loop-the-loop of radius r, as shown above. What is the approximate minimum velocity required to achieve this goal?

(A) $(gr)^{1/2}$
(B) $(5gr)^{1/2}$
(C) $2(gr)^{1/2}$
(D) $(2gr)^{1/2}$

QUESTIONS 23–25 ARE BASED ON THE FOLLOWING CIRCUIT:

23. What is the equivalent resistance of the circuit?

(A) 6 Ω
(B) 8 Ω
(C) 11 Ω
(D) 5 Ω

24. What is the reading of the voltmeter across the 4-ohm resistor?

(A) 3 V
(B) 9 V
(C) 12 V
(D) 6 V

25. What is the current through the leftmost 2-ohm resistor?

(A) 3 A
(B) 1.5 A
(C) 0.75 A
(D) 1 A

26. A wire segment in a circuit with a cross-sectional area of A and length L is replaced by a wire segment made of the same material that has twice the area but half the length. The resistance of the new segment, compared to the original segment, will

(A) be reduced by half

(B) be reduced to one-quarter

(C) quadruple

(D) remain the same

27. In the fixed standing wave shown below (imagine a string of fixed length), what will happen to the wavelength and frequency if the wave speed is raised while the standing wave pictured remains unchanged?

(A) Wavelength increases, while frequency remains the same.

(B) Wavelength remains the same, while frequency increases.

(C) Both wavelength and frequency increase.

(D) Although we know the product of wavelength and frequency increases, we do not know what combination is producing this effect.

28. Bats can find objects in the dark by using echolocation (sending out a high-frequency sound and listening to the echo). They also listen for a change in the pitch of the echo to

(A) confirm the estimated distance of the object

(B) determine the velocity of the object

(C) determine whether the object is moving toward or away from them

(D) estimate the approximate composition of the object

29. A 340-hertz tuning fork sets an air column vibrating in fundamental resonance, as shown above. A hollow tube is inserted into a column of water, and the height of the tube is adjusted until strong resonance is heard. At approximately what length of the air column will this happen?

(A) 100 cm

(B) 75 cm

(C) 50 cm

(D) 25 cm

30. How many different directions can a two-dimensional vector have if its components are of equal magnitude?

    (A) One, at 45°
    (B) Two
    (C) Four
    (D) Infinite

31. Two vectors, $\vec{A}$ and $\vec{B}$, have components $A_x = -2$, $A_y = 3$, $B_x = 5$, and $B_y = 1$. What is the approximate magnitude of the vector $\vec{A} + \vec{B}$?

    (A) 3
    (B) 4
    (C) 5
    (D) 7

32. As the angle between two vectors increases from 0° to 180°, the magnitude of their resultant

    (A) increases, only
    (B) increases and then decreases
    (C) decreases, only
    (D) decreases and then increases

33. At what angle should a projectile be launched in order to achieve the maximum range for a given initial velocity under no air resistance?

    (A) 90°
    (B) 30°
    (C) 45°
    (D) 60°

34. An object is dropped from a height of 45 m. Neglecting air resistance, what is the approximate velocity of the object as it hits the ground?

    (A) 10 m/s
    (B) 15 m/s
    (C) 20 m/s
    (D) 30 m/s

35. A boat moving due north crosses a river 240 meters wide with a velocity of 8 meters per second relative to the water. The river flows east with a velocity of 6 meters per second. How far downstream will the boat be when it has crossed the river?

    (A) 240 m
    (B) 180 m
    (C) 420 m
    (D) 300 m

A variable force acts on a 2-kilogram mass according to the graph below.

36. How much work was done while displacing the mass 10 meters?

    (A) 40 J

    (B) 38 J

    (C) 32 J

    (D) 30 J

37. What was the average force supplied to the mass for the entire 10-meter displacement?

    (A) 3.2 N

    (B) 1.2 N

    (C) 4.4 N

    (D) 4 N

38. A man weighing himself is standing on a bathroom scale in an elevator that is accelerating upward at a rate of 0.5 meter per second squared. By what percentage is the reading of the scale off from the person's true weight?

    (A) 0% (accurate)

    (B) 5% too high

    (C) 5% too low

    (D) 0.5% too high

39. A 50-kilogram person is sitting on a seesaw 1.2 meters from the balance point. On the other side, a 70-kilogram person is balanced. How far from the balance point is the second person sitting?

    (A) 0.57 m

    (B) 0.75 m

    (C) 0.63 m

    (D) 0.86 m

40. An object rolls down a steep incline with very little friction. At the same time, an object of equal mass slides down a similar but frictionless incline. Which one takes less time to get to the bottom?

    (A) They take the same time.
    (B) The rolling object takes less time.
    (C) The sliding object takes less time.
    (D) The answer depends on the rolling object's moment of inertia.

41. What is the value of g at a position above Earth's surface equal to Earth's radius?

    (A) 9.8 N/kg
    (B) 4.9 N/kg
    (C) 2.45 N/kg
    (D) 1.6 N/kg

42. If it takes a minimum of 5 N to move 2 C of charge away from some other charge, how much energy is needed to move 3 C of charge 40 cm in the same situation?

    (A) 2.5 J
    (B) 7.5 J
    (C) 2.0 J
    (D) 3.0 J

43. If an object is spinning at 150 RPM (revolutions per minute) and comes to stop in 2 seconds, what is its average acceleration in radians/s^2?

    (A) -2.5π
    (B) -2π
    (C) -5π
    (D) -10π

44. What is the work done by a horizontal spring (spring constant k) expanding from a compression distance x to an extension distance x to an attached mass?

    (A) $2kx^2$
    (B) $\frac{1}{2}kx^2$
    (C) kx^2
    (D) 0

45. If an isolated spinning object's moment of inertia is reduced by a factor of 3 by internal forces, how will its angular momentum change?

    (A) Angular momentum will be 3 times its previous value.
    (B) Angular momentum will be reduced to 1/3 its previous value.
    (C) Angular momentum will be reduced to 1/9 its previous value.
    (D) Angular momentum will remain unchanged.

46. If the tension in a taut string is increased, which of the following will also be increased when the fundamental frequency is struck? Select two correct answers.

    (A) The velocity of propagation
    (B) The frequency of the fundamental
    (C) The wavelength of the fundamental
    (D) The amplitude of the wave

47. A friend is balancing a fork on one finger. Which of the following are correct explanations of how he accomplishes this? Select two correct answers.

   (A) Total energy is conserved.
   (B) The fork's moment of inertia is zero.
   (C) The fork's center of mass is above his finger.
   (D) The fork's clockwise torque is equal to its counterclockwise torque.

48. Two tuning forks are struck at the same time. A beat frequency of 12 beats per second is observed. If one tuning fork has a frequency of 384 Hz, what could be the frequency of the second tuning fork? Select two correct answers.

   (A) 260 Hz
   (B) 372 Hz
   (C) 396 Hz
   (D) 408 Hz

49. A book rests on top of a table. Which of the following are action-reaction pairs described by Newton's third law? Select two correct answers.

   (A) The weight of the book and the normal force of the table upon the book
   (B) The weight of the book and the weight of the table
   (C) The weight of the table and the upward pull of the table on Earth
   (D) The normal force upon the book from the table and the downward force on the table due to the book

50. Consider the impulse received by the first car in each of the following cases. In each case, the cars are at rest after the collision. In which two of the following cases will the car receive the same impulse? Select two correct answers.

   (A) A 5,000-kg car traveling at 10 m/s has a head-on collision with an equal and oppositely directed second car.
   (B) A 5,000-kg car traveling at 10 m/s has a head-on collision with a large building.
   (C) A 2,500-kg car traveling at 10 m/s has a head-on collision with an equal and oppositely directed second car.
   (D) A 2,500-kg car traveling at 5 m/s has a head-on collision with an equal and oppositely directed second car.

SECTION II: FREE-RESPONSE

Time: 90 minutes

5 questions

> **DIRECTIONS:** You have 90 minutes to complete this portion of the test. You may use a calculator and the information sheets provided in the appendix.

1. (7 points; ~13 minutes) Use the sketch to answer the following questions.

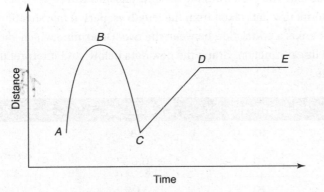

(a) Use the qualitative keywords below to fill in the chart beneath with a brief description of the motion:

List 1: constant, increasing, decreasing

List 2: negative, positive, zero

Time Interval	Velocity (One Word from List 2)	Speed (One Word from List 1)	Acceleration (One Word from List 2)
$A \rightarrow B$	pos	~~const~~ inc.	~~0~~ pos
$B \rightarrow C$	~~neg~~ pos	decrease	neg
$C \rightarrow D$	pos	inc	0
$D \rightarrow E$	pos	const	0

(b) Which is faster, the average velocity from C to D or the average velocity from C to E? Explain your reasoning. C-D $\frac{D}{T} = V$ C-E+7 C-D+

(c) What is the instantaneous velocity at point B? Explain your reasoning. O because top of curve

(d) Pick a couple of points on this graph that are not a realistic representation of a real-world object's motion. Explain what is problematic about these two points. C & D because things don't instantly speed up or stop

(e) Sketch a velocity versus time graph from these data. Label points A, B, C, D, and E.

V

2. (12 points; ~25 minutes) A group of students are given the following supplies: a stopwatch, a long string, various metersticks and protractors, and a large supply of various styles of predetermined masses.

(a) Describe three short experimental procedures to determine the dependency of a simple pendulum's period of oscillation on amplitude, mass, and length. You may include a labeled diagram of your setup to help in your description. Indicate what measurements you would take and how you would take them. Include enough detail so that another student could carry out your procedure.

(b) What are the expected results of each investigation? Sketch out what the data will look like in each of the three investigations (amplitude, mass, and length).

(c) What are the common sources of error or expected deviations from ideal results that might happen during this investigation? Which of the three investigations might you expect to deviate the most from the ideal results and why?

(d) Here are some raw data taken from the length vs. period investigation by a student who suspects there is a correlation between the two. Determine which variables to graph to produce a linear function. Graph the new data below, and interpret the slope of your best-fit line.

Length (cm)	Period (s)	?
10	0.62	
20	0.90	
30	1.09	
40	1.28	
55	1.48	
75	1.75	
85	1.85	

3. (12 points; ~25 minutes) A mass m is resting at a height h above the ground. When released, the mass can slide down a frictionless track to a loop-the-loop of radius r as shown below.

(a) As the mass slides down the incline, it gains speed. However, the mass may or may not "make it" around the loop without falling out. Why and under what conditions will the mass travel around the loop without falling out? Why and under what conditions will it not complete the loop? Explain your reasoning qualitatively, making sure you address the normal force experienced by the mass.

(b) Two fellow students are arguing about whether the maximum possible r of a loop that the mass can make it around will be linearly dependent on height h or if the relationship will be some power law. Without solving for the relationship explicitly, indicate a line of reasoning that would settle this argument.

(c) Within the loop, the normal force is always acting in a centripetal direction. Explain why the normal force is not always equal to mv^2/r at each point. Are there any points along the loop where the normal force is equal to mv^2/r? Support your argument with free-body diagrams.

(d) Using quantitative reasoning, derive an expression for the normal force in terms of mg when the mass is halfway up the loop. Assume that the minimum height, h, that enables a complete loop has been chosen.

(e) If the minimum height h has been found for a fixed loop of size r, but a rolling ball of mass m is substituted for the sliding mass, the ball will (indicate your choice):

_____ make it through the loop more easily

_____ pass through the loop the same as before

_____ fall out of the loop

Justify your answer qualitatively, with no equations or calculations.

4. (7 points; ~13 minutes) A 2-kilogram mass is twirled in a vertical circle as shown. It is attached to a 2-meter rope. As the mass just clears the ground (point A), its velocity is 10 m/s to the right. Neglect any air resistance. When the mass reaches point B, it makes a 30-degree angle to the horizontal as shown.

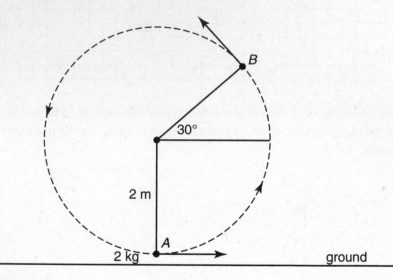

(a) Describe qualitatively, using words and diagrams, the difference in the object's speed at point B if the twirling is done (i) at constant speed or (ii) at constant tension. Which would result in the greater speed at point B? Include an analysis of how the tension must change in part (i).

(b) At point B, the rope is released and the mass becomes a projectile. Assume the rope has a uniform distribution of mass and remains attached to the flying object. Which of the following options best describes the projectile's range as compared with a launch without the rope?

_____ addition of rope makes no difference to the range

_____ addition of rope increases the range

_____ addition of rope decreases the range

Explain your reasoning.

(c) When twirling the object at constant speed, explain in terms of work and energy how the constant speed is maintained. Is the path truly circular when the object's speed is maintained? In your explanation, be explicit about which forces are doing positive work and which are doing negative work. Your answer should be in a clear, coherent paragraph.

5. (7 points; ~13 minutes) You are given three 2-ohm resistors, some wire, a variable DC voltage supply, a voltmeter, and an ammeter.

(a) Draw a schematic diagram of a circuit that will produce an equivalent resistance of 3 ohms, including ammeter and voltmeter, as appropriate, to be able to measure the circuit current and source voltage.

(b) Which, if any, of these components are assumed to have zero internal resistance? Which, if any, of these components are assumed to have infinite resistance? Justify your choices.

(c) Trace out one complete loop in your circuit, and prove Kirchhoff's loop rule. Once you set up your rule, assume a setting of 6 volts on the power supply. Verify your rule numerically.

(d) A student wishes to measure the voltage and current in a simple circuit using small lightbulbs instead of commercially manufactured resistors. She finds that after a short while, the current in the ammeter is decreasing. How might she account for this?

ANSWER KEY

1. **D**		14. **A**		27. **B**		40. **C**
2. **C**		15. **C**		28. **C**		41. **C**
3. **D**		16. **B**		29. **D**		42. **D**
4. **B**		17. **C**		30. **C**		43. **A**
5. **B**		18. **D**		31. **C**		44. **D**
6. **C**		19. **D**		32. **C**		45. **D**
7. **C**		20. **A**		33. **C**		46 **A, B**
8. **B**		21. **B**		34. **D**		47. **C, D**
9. **B**		22. **B**		35. **B**		48. **B, C**
10. **A**		23. **D**		36. **C**		49. **C, D**
11. **D**		24. **D**		37. **A**		50. **A, B**
12. **D**		25. **B**		38. **B**		
13. **A**		26. **B**		39. **D**		

ANSWERS EXPLAINED

Section I: Multiple-Choice

1. **(D)** Twice the initial vertical velocity will give twice the time in flight. Average vertical velocity will also be doubled. Displacement is the product of these two: $2 \times 2 = 4$.

   Alternatively, one could use $v^2 = v_0^2 + 2ad$, since v at the maximum height is zero. Since v_0 is doubled and then squared, the vertical displacement d must be 4 times bigger.

   Free-fall problems are independent of mass.

2. **(C)** Conservation of momentum:

$$p_{tot} = 2(+4) + 5(-1) = +3 \text{ kg m/s} = mv = (7 \text{ kg})v$$
$$v = +3/7 \text{ m/s}$$

   The collision is inelastic since the carts stick together.

3. **(D)** Higher altitude is strictly a function of V_Y (Projectile Y). Range is a function of both V_X and V_Y such that the angles equally above and below 45 degrees (the max angle for range) will result in equal horizontal displacement. Note that 30 degrees and 60 degrees are both 15 degrees off from 45 degrees.

4. **(B)** Since the 10 N force will start the mass moving, this must be greater than the static friction force holding the mass in place. The maximum *static* friction must be 10 N or less.

$$F_s < 10 \text{ N}$$
$$\mu N < 10 \text{ N}$$
$$\mu mg < 10 \text{ N}$$
$$\mu (5)(10) < 10$$
$$\mu < 1/5$$

5. **(B)** The object is moving, so the velocity is not zero. The object is not accelerating, so velocity is constant.

6. **(C)** In general:

$$y(t) = y_0 - v_{y0} - \frac{1}{2} g t^2$$
$$x(t) = v_{x0} t$$

Solve for t in the last equation. Then plug back into the first equation and substitute in $y_0 = h$, $v_{y0} = 0$, and $v_{x0} = v_0$:

$$y(x) = h - (0)(x/v_0) - \frac{1}{2} g (x/v_0)^2 = h - g x^2 / (2 v_0^2)$$

7. **(C)** Maximum acceleration allows you to determine maximum displacement:

$$ma = kA$$

Knowing the amplitude allows you to determine easily the maximum speed via energy conservation:

$$\frac{1}{2} k A^2 = \frac{1}{2} m v^2$$

8. **(B)** Maximum velocities happen when going through the equilibrium point (zero acceleration and zero displacement): all kinetic energy (KE) and no potential energy (PE).

9. **(B)** Momentum is conserved during the collision, which enables us to solve for an initial upward velocity of the combination. Then energy conservation can be used to relate the height to that initial upward velocity. Choice (A) cannot be used because some unknown amount of mechanical energy will be lost by the bullet embedding itself in the wood.

10. **(A)** Constant velocity means no acceleration, so $\vec{F}_{net} = 0$.

$$F_{push} - F_{friction} = 0$$
$$F_{push} - \mu_k N = F_{push} - \mu_k mg = 0$$
$$F_{push} = \mu_k mg = (0.1)(10 \text{ kg})(10 \text{ m/s}^2) = 10 \text{N}$$

11. **(D)** First find the acceleration of the system:

$$\vec{F}_{net} = m\vec{a}$$
$$10\,N = (4+1\,kg)a$$
$$a = 2\,m/s^2$$

The contact force, P, is the only force felt by the 1-kg mass:

$$\vec{F}_{net} = m\vec{a}$$
$$P = (1\,kg)(2\,m/s^2) = 2\,N$$

12. **(D)** $P = work/t = Fd/t = mgh/t$

13. **(A)** Impulse can be found by Ft if the details of the force are known or, alternatively, by:

$$Impulse = \Delta p = (1{,}500\,kg)(0\,m/s - 25\,m/s) = -37{,}500\,N \cdot s$$

Ignore the minus sign as the question asks about magnitude.

14. **(A)** When the block is at the maximum height, static friction is obtained:

$$\vec{F}_{net} = mg\,\sin\theta - \mu N = 0$$
$$mg\,\sin\theta - \mu(mg\,\cos\theta) = 0$$
$$\mu = (\sin\theta)/(\cos\theta) = \tan\theta$$

15. **(C)** Doubling the period requires a quadrupling of length:

$$T = 2\pi(l/g)^{\frac{1}{2}}$$

16. **(B)** Impulse = area of F vs. T graph = $\Delta p = mv_f - 0$

Solving for v_f = (area under graph)$/m$

17. **(C)** Conservation of energy:

$$\frac{1}{2}mv^2 = mg\Delta h$$
$$\frac{1}{2}v^2 = g\Delta h = (10\,m/s^2)(2.25\,m - 0.75\,m)$$
$$v = (30)^{\frac{1}{2}} = 5.5\,m/s$$

18. **(D)** Gravitational field strength halfway between any two equal masses is always zero as each contributes oppositely directed gravitational fields.

19. **(D)** Conservation of energy while on the frictionless hill can give the speed at the top of the 10-m hill:

$$\frac{1}{2}mv^2 = KE = mg\Delta h$$

Then impulse equals change in momentum can be used since the final momentum must be zero.

20. **(A)**

$$\text{Mechanical energy} = mgh + \frac{1}{2}mv^2$$

$$\text{Energy "lost" or "gained"} = mgH_2 + \frac{1}{2}mv_2^2 - \left(mgH_1 + \frac{1}{2}mv_1^2\right)$$

Work by forces other than gravity = change in energy

21. **(B)** Energy conservation:

$$\frac{1}{2}kx^2 = mgh$$

Solving for h:

$$h = kx^2/2mg$$

Doubling x will quadruple the height, whereas all other factors will only double the height.

22. **(B)** Circular motion at the top:

$$F_{\text{net}} = mg + N = mv^2/r$$

The lowest speed will be when there is no normal force ($N = 0$):

$$mg = mv^2/r$$

$$v_{\text{top}} = (gr)^{\frac{1}{2}}$$

Kinetic energy at the bottom must give both this speed and potential energy to gain $2r$ in height:

$$E_{\text{bottom}} = E_{\text{top}}$$

$$\frac{1}{2}mv^2 = gm(2r) + \frac{1}{2}mv_{\text{top}}^2$$

$$\frac{1}{2}mv^2 = gm(2r) + \frac{1}{2}m(gr)$$

Solving for v:

$$v = \left(5gr\right)^{\frac{1}{2}}$$

23. **(D)** The two 2-Ω resistors are in series with each other: $2\ \Omega + 2\ \Omega = 4\ \Omega$.

This 4-Ω equivalent resistance is in parallel with the existing 4-Ω resistor: $\frac{1}{4}\ \Omega + \frac{1}{4}\ \Omega = \frac{1}{2}\ \Omega$. So the equivalent resistance for that section of the circuit is $2\ \Omega$. This 2-Ω equivalent resistance is in series with the 3-Ω resistor: $2\ \Omega + 3\ \Omega = 5\ \Omega$.

24. **(D)** The 15 volts must be split between the 3-ohm resistor and the 2-ohm equivalent resistance of the right-hand side (see answer 23). Voltage for the 4-ohm resistor is the same as the voltage across the 2-ohm equivalent resistance of the right-hand side:

$$V = (2/5)(15\ \text{V}) = 6\ \text{V}$$

25. **(B)** The 15-volt battery will supply 3 amps of current for the 5-ohm circuit:

$$V = IR$$
$$15\ \text{V} = (3\text{A})(5\ \text{ohms})$$

This 3-amp current will split as it comes to the branching point before the 4-ohm resistor. Since both pathways have equal resistance (4 ohms), the current will split evenly: 1.5 amps.

26. **(B)** Resistance is determined by:

$$\rho L/A$$

The same material means the resistivity, ρ, is the same. Letting $L \rightarrow L/2$ and $A \rightarrow 2A$:

$$\rho L/A \rightarrow \rho(L/2)/(2A) = (\rho L/A)/4$$

27. **(B)** The wavelength is set by the length of the standing wave and the number of the harmonic. Therefore, the wavelength remains the same. The frequency of the wave must have been raised to correspond to the higher wave speed: $\lambda f = v$.

28. **(C)** The Doppler effect can be used to determine relative speed toward or away from the source by looking at frequency shift in the reflected wave. Note that the velocity vector cannot be determined. Components of velocity not directed toward or away from the receiver do not contribute to the Doppler shift.

29. **(D)** The fundamental frequency in a half-pipe is $\frac{1}{4}$ wavelength:

$$\text{Wavelength} = \text{wave speed/frequency} = 340/340 = 1\ \text{m}$$

The pipe needs to be $\frac{1}{4}$ of this: 0.25 m = 25 cm.

30. **(C)** Any 45-45-90 triangle has sides of equal length. Therefore, a 45° vector would have equal length components. So would 135°, 225°, and 315°, as these vectors all form 45-45-90 triangles with their components.

31. **(C)** Add the components of the vectors:

$$A_x + B_x = -2 + 5 = 3$$
$$A_y + B_y = 3 + 1 = 4$$

Use the Pythagorean theorem on these net components:

$$3^2 + 4^2 = 5^2$$

32. **(C)** Maximum vector addition is when the two vectors are aligned ($0°$) and minimum is when they are opposite ($180°$). All values in between will steadily decrease the resultant from maximum to minimum.

33. **(C)** Maximum range is at $45°$ since this gives decent time in flight and decent horizontal velocity.

34. **(D)** First, find the time in flight:

$$h = \frac{1}{2}(10)t^2$$
$$45 = 5t^2$$
$$t = 3 \text{ s}$$

Now use this time to find the final speed:

$$v_f = at = (10 \text{ m/s}^2)(3 \text{ s}) = 30 \text{ m/s}$$

35. **(B)** Find time to cross the river using only perpendicular components:

$$D = V_y t$$
$$240 = 8t$$
$$t = 30 \text{ s}$$

Next find the downstream distance using the parallel component:

$$D = V_x t$$
$$D = (6 \text{ m/s})(30 \text{ s}) = 180 \text{ m}$$

36. **(C)** Work is the area under the curve:

$$(4 \text{ N})(6 \text{ m}) + (2 \text{ N})(4 \text{ m}) = 24 \text{ J} + 8 \text{ J} = 32 \text{ J}$$

37. **(A)** work $= F_{average} \cdot$ displacement

$$32 \text{ J} = (F_{average})(10 \text{ m})$$
$$F_{average} = 3.2 \text{ N}$$

38. **(B)** The bathroom scale reads the normal force:

$$F_{net} = N - mg = ma$$
$$0.5 \text{ m/s}^2 = 0.05g$$
$$N - mg = m(0.05g)$$
$$N = m(1.05g)$$

The reading will be 5% too high.

39. **(D)** Balanced means the torques are equal and opposite:

$$50 \text{ kg} \cdot g \cdot 1.2 = 70 \text{ kg} \cdot g \cdot x$$
$$x = 6/7 = 0.86 \text{ m}$$

40. **(C)** Both objects start with the same potential energy. However, the rolling object must use some of that potential energy for rotational energy, leaving less for linear kinetic energy. Therefore, the rolling object moves more slowly down the hill.

41. **(C)** Universal gravity (and the gravitational field) are $1/r^2$ laws; doubling r will quarter the field. $(1/4)g = (1/4)(9.8 \text{ N/kg})$ Note that $\text{N/kg} = \text{m/s}^2$.

42. **(D)** First, both situations will have the same force to charge ratio:

$$F/q = 5/2 \text{ N/C}$$

Next, find the force for the new charge:

$$(5/2)(3) = 7.5 \text{ N}$$

Now, find the work done:

$$W = F \cdot D = (7.5 \text{ N})(0.4 \text{ m}) = 3 \text{ J}$$

43. **(A)** First, convert RPMs to rad/s:

$$150 \text{ rev/min} \times (2\pi/1 \text{ rev}) \times (1 \text{ min}/60 \text{ s}) = 5\pi \text{ rad/s}$$

$$\text{acceleration} = \text{change in velocity/time} = (0 - 5\pi \text{ rad/s})/(2 \text{ s}) = -2.5\pi \text{ rad/s}^2$$

44. **(D)** $\frac{1}{2}kx^2$ of work is delivered to the mass while uncompressing, followed by $-\frac{1}{2}kx^2$ done while the mass extends the spring outward, totaling to 0 net work for the entire expansion. Alternatively, think about the velocity being zero at the beginning and end of that single oscillation. No change in kinetic energy occurs; therefore, no net work is done.

45. **(D)** No external torque means momentum must be conserved.

46. **(A) and (B)** Higher tension means waves will propagate at higher speeds. Since the wavelength of the standing wave is fixed by the length of the string, the frequency of the fundamental must also increase.

47. **(C) and (D)** An object's center of mass is the balancing point by definition because the moment arm on one side is the same as on the other. Therefore, the torque caused by gravity on either side will cancel.

48. **(B) and (C)** The beat frequency is caused by the difference between the two frequencies: $384 \pm 12 = 396$ or 372 Hz.

49. **(C) and (D)** Action-reaction pairs must be the same type of force and found on each partner of the force exchanging objects. The weight of an object is caused by the entire planet. Therefore, that reaction force is on the planet itself. Likewise, the normal forces exchanged between the bottom surface of the book and top surface of the table are an action-reaction pair.

50. **(A) and (B)** The cars will both experience an impulse of $50,000$ N · s. The other two cases result in smaller impulses.

Section II: Free-Response

1. (a) (3 points; 1 point for each column in the table)

Time Interval	Velocity (One Word from List 2)	Speed (One Word from List 1)	Acceleration (One Word from List 2)
$A \rightarrow B$	positive	decreasing	negative
$B \rightarrow C$	negative	increasing	negative
$C \rightarrow D$	positive	constant	zero
$D \rightarrow E$	zero	constant	zero

(b) (1 point) The slope from C to D is steeper than the slope drawn from C to E. Therefore, the average velocity is greater from C to D.

(c) (1 point) Point B is a turnaround point. The slope of a tangent line drawn at that point would be horizontal. For these reasons, the instantaneous velocity at this point is zero.

(d) (1 point) Points C and D are not continuous, which means the slope at these points is undefined. Therefore, the instantaneous velocity at these two points does not exist. Real-world motion is always continuous.

(e) (1 point)

2. (a) (3 points) This is really three separate investigations. In each case, students should measure the period repeatedly and then take the mean value.

1. (1 point) Amplitude versus period. Take a period measurement for 5 to 10 different amplitudes while keeping the mass and length the same. The amplitude can be controlled by pulling the string-mass combo out to a certain angle as measured by the protractor.
2. (1 point) Mass versus period. Take a period measurement for 5 to 10 different masses while keeping the amplitude and length constant.
3. (1 point) Length versus period. Take a period measurement for 5 to 10 different lengths while keeping the mass and amplitude constant. Length is measured from the pivot point to the center of mass.

(b) (3 points; 1 point for each graph) No correlation is expected for amplitude and mass variations. A nonlinear relationship is expected for length and period:

$$T_p = 2\pi(L/g)^{\frac{1}{2}}$$

(c) (2 points) Beyond the usual random errors of measurement (especially when using a timer but minimized by taking the median value of a few trials each time), one predictable deviation is in the amplitude investigation. Pendulums actually behave as simple harmonic oscillators only under the conditions of small angles (small enough that $\sin \theta$ is approximately θ). A large enough amplitude will require the pendulum to oscillate at larger angles. This means that at large amplitudes, one can predict the results to deviate from the expected as the gravitational force no longer acts as a simple restorative force. (1 point)

One other possible source of systematic error would be in the mass investigation. As various masses are swapped out on a fixed length of string, the students may inadvertently be changing the length of the string when adding on different-sized masses. The length of the pendulum is from the pivot point to the *center of mass*. If the students did not compensate for this when adding larger masses by shortening the string, they may see an artificial relationship at higher masses in their graph of mass versus period. (1 point)

(d) (4 points) To obtain a linear graph, one must plot T^2 versus L:

$$T^2 = (4\pi^2/g)L$$

The slope of this plot would then be equal to $4\pi^2/g$. Alternatively, one could plot T versus $\sqrt{L}$. (1 point for the table; 2 points for the graph)

Length (cm)	Period (s)	T^2 (s^2)
10	0.62	0.38
20	0.90	0.81
30	1.09	1.19
40	1.28	1.64
55	1.48	2.19
75	1.75	3.06
85	1.85	3.42

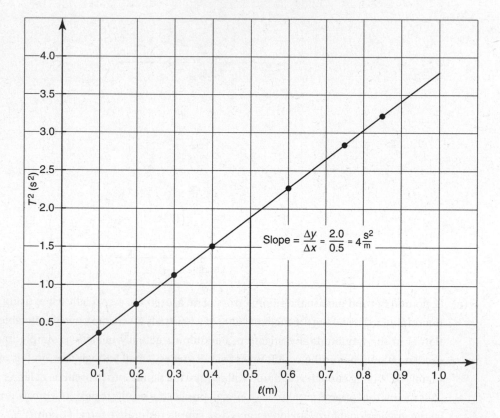

The measured slope is 4 s^2/m. It can be compared to the theoretical prediction of slope:

$$4\pi^2/g = 4.02 \text{ s}^2/\text{m}$$

(1 point)

3. (a) (2 points) As the value h is raised, the mass gains velocity at the bottom of the incline. As the mass slides around the inside of the loop, it will lose speed as it goes back up. The entire time the mass is inside the loop, it must be undergoing centripetal acceleration (specifically, its v^2/r must equal its inward acceleration). If the mass is moving too slowly to maintain its circular motion, it will fail to complete the loop. The centripetal force is supplied by the normal force throughout the trajectory (sometimes also with or against the force of gravity). If the normal force goes to zero, this indicates a loss of contact between the sliding mass and the track. The slowest speed will occur at the top of the loop. Therefore, this is the most likely place for the mass to fall. At the top, gravity is acting centripetally along with the normal force, so Newton's second law gives us:

$$N + mg = mv^2/r \qquad \text{(1 point)}$$

As h gets smaller, the speed at the top will get smaller, which will decrease the right-hand side of the equation above. At some point, mv^2/r will get as small as mg, at which point the normal force will be at zero. Any speed greater than this will let the mass complete the loop. Any speed lower than this will cause the mass to fail to complete the loop. (1 point)

(b) (2 points) The object has gravitational potential energy in the beginning, which is linearly proportional to h. Part of the potential energy is transformed into kinetic energy at the top of the loop. Kinetic energy is proportional to v^2. However, centripetal force is also proportional to v^2, and the centripetal force is linear with respect to the normal force. It is this normal force going to zero that determines the critical speed. So the student who claims there is a linear dependence between h and r is correct.

(c) (3 points) Because the normal force is not always the only centripetal force, it alone is not equal to mv^2/r. Generally speaking, a component of the force due to gravity is acting centripetally (top half of the loop) or centrifugally (bottom half of the loop). However, at the points halfway between the top and bottom, the normal force indeed is acting alone centripetally and is equal to mv^2/r.

The following drawings show only the radial component of the two forces (F_N, F_g).

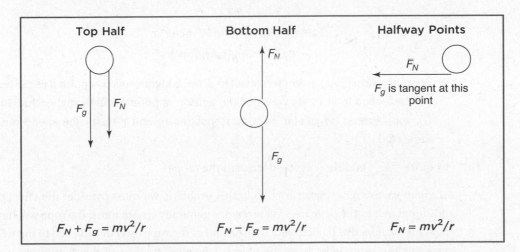

Tangent components of gravity speed up or slow down the object but do not participate in the centripetal acceleration.

(d) (3 points) First, determine the velocity at the top of the loop by setting $N = 0$ in the expression for net force at the top of the loop from part (a):

$$v_{top}^2 = gr \qquad \text{(1 point)}$$

The mechanical energy is conserved. Set the expression at the top of the loop equal to that at midpoint:

$$mg(2r) + \frac{1}{2}mv_{top}^2 = mg(r) + \frac{1}{2}mv^2$$

Substitute in our expression for v_{top}:

$$mg(2r) + \frac{1}{2}m(gr) = mgr + \frac{1}{2}mv^2$$

Solving the above for the velocity at the halfway point:

$$v^2 = 3gr \qquad \text{(1 point)}$$

Using our expression from part (c), the normal force is the centripetal force at this position:

$$F_N = \frac{mv^2}{r} = m \times \frac{3gr}{r} = 3mg \qquad \text{(1 point)}$$

(e) (2 points) __X__ fall out of the loop

Since the original gravitational potential energy must now be going into rotational kinetic energy, less energy will be available for the tangential speed needed to make it through the loop. Since the height, h, was set minimally for the sliding block, the rolling object will not have enough linear kinetic energy for its tangential speed to clear the top of the loop.

4. (a) (4 points)

(i) (2 points) To maintain constant speed, the tension must be constantly adjusted as the object undergoes its centripetal motion. The tension will be lower at point B than it is at the bottom as there is a component of gravity "helping" the tension in the centripetal direction (as opposed to gravity acting centrifugally at point A):

$$F_{cA} = F_{cB} = mv^2/r$$
$$T_A - mg = T_B + mg\sin 30° = mv^2/r$$
$$T_B = T_A - mg(1 + \sin 30°)$$

(ii) (2 points) Constant tension will result in a much higher speed at point B as the tension will be locked in at T_A above. Since the tension at point B will be higher than in case (i), mv^2/r must be greater as well. Constants m and r mean the speed must be increased.

(b) (1 point) __X__ addition of rope decreases the range

Although we are ignoring air friction in this problem, we must consider the effect of the additional mass. Although free-fall problems generally ignore mass, the rope will have an effect on the effective launch velocity. The end of the rope directly attached to the twirling mass has the same velocity as the object. However, the rest of the distributed mass is moving at a lower velocity (down to zero at the center of the circle). Therefore, the net effect of the added mass will be to decrease the launch velocity. In addition, recall that the equations of motion actually track the center of mass of an object. Since the center of mass for the combined rope-plus-object system is lower (the mass of the rope extends below the 2-kg mass), the combined rope-plus-object system is effectively being launched from a lower height. A lower launch velocity from a lower height will result in a shorter range for the projectile.

(c) (2 points) While on the way up during the circular motion, the force due to gravity is doing negative work. The component of weight that is tangent does this negative work. Therefore, to maintain the same kinetic energy, the tension must do some positive work. In order to accomplish this, the tension must supply an upward tangent component as well; that is, the circle cannot be perfect. On the downward half of the circular motion, the positive and negative work roles of the tangent components of gravity and tension must reverse.

5. (a) (2 points)

(b) (2 points) Ammeters, wires, and power supplies are assumed to have zero internal resistance as they have no resistance indicated in the circuit diagram. If there were any resistance, unaccounted losses of voltage would occur in our circuit. When modeling actual resistance inside of components, an "internal" resistor is added to the circuit diagram. (1 point)

Voltmeters would create a parallel path if they did not have infinite resistance. To the extent they have less than infinite resistance, the addition of voltmeters to the circuit would effectively lower the resistance, thereby lowering the voltage reading. (1 point)

(c) (2 points) Kirchhoff's loop rule states that the sum of voltages changing around the loop must be zero. As an example of this, start to the right of the power supply and trace a loop clockwise through the top parallel resistor and then the resistor to the right and back to our starting point. Tracing our changes in voltage:

$$+V - (I/2)R - IR = 0 \qquad \text{(1 point)}$$

For the entire circuit:

$$V = IR$$
$$6 = I(3)$$
$$I = 2 \text{ amps} \qquad \text{(1 point)}$$

Loop rule:

$$+6 - (1)(2) - (2)(2) = 6 - 2 - 4 = 0$$

(d) (1 point) The lightbulbs get hot very quickly, which increases their electrical resistance. This, in turn, reduces the current measured by the ammeter.

TEST ANALYSIS

PRACTICE TEST 1

Section 1: Multiple-Choice

Note that the questions requiring two answers are to be graded as completely correct (1 point) or incorrect (0 points, even if you have one of the two answers correct).

$$\text{Number correct (out of 50)} = \overline{\qquad\text{Multiple-Choice Score}\qquad}$$

Section 2: Free-Response

Partial credit is awarded for any correct responses within an individual free-response question.

$$\text{Question 1} = \overline{\qquad\qquad\qquad}_{\text{(out of 7)}}$$

$$\text{Question 2} = \overline{\qquad\qquad\qquad}_{\text{(out of 12)}}$$

$$\text{Question 3} = \overline{\qquad\qquad\qquad}_{\text{(out of 12)}}$$

$$\text{Question 4} = \overline{\qquad\qquad\qquad}_{\text{(out of 7)}}$$

$$\text{Question 5} = \overline{\qquad\qquad\qquad}_{\text{(out of 7)}}$$

$$\text{Total} = \overline{\qquad\qquad\qquad}_{\text{(out of 45)}}$$

Final Score

$$\overline{\qquad\text{Multiple-Choice Score}\qquad} + \left(1.11 \times \overline{\qquad\text{Free-Response Score}\qquad}\right) = \overline{\qquad\text{Total (out of 100)}\qquad}$$

Final Score Range*

Final Score Range	AP Score
71–100	5
56–70	4
41–55	3
26–40	2
0–25	1

*Note: The guidelines above are based on the released scores for past AP Physics 1 exams. Actual score ranges vary from year to year and are determined by the College Board each year. Thus, the ranges shown are approximate.

ANSWER SHEET
Practice Test 2

1. Ⓐ Ⓑ Ⓒ Ⓓ
2. Ⓐ Ⓑ Ⓒ Ⓓ
3. Ⓐ Ⓑ Ⓒ Ⓓ
4. Ⓐ Ⓑ Ⓒ Ⓓ
5. Ⓐ Ⓑ Ⓒ Ⓓ
6. Ⓐ Ⓑ Ⓒ Ⓓ
7. Ⓐ Ⓑ Ⓒ Ⓓ
8. Ⓐ Ⓑ Ⓒ Ⓓ
9. Ⓐ Ⓑ Ⓒ Ⓓ
10. Ⓐ Ⓑ Ⓒ Ⓓ
11. Ⓐ Ⓑ Ⓒ Ⓓ
12. Ⓐ Ⓑ Ⓒ Ⓓ
13. Ⓐ Ⓑ Ⓒ Ⓓ

14. Ⓐ Ⓑ Ⓒ Ⓓ
15. Ⓐ Ⓑ Ⓒ Ⓓ
16. Ⓐ Ⓑ Ⓒ Ⓓ
17. Ⓐ Ⓑ Ⓒ Ⓓ
18. Ⓐ Ⓑ Ⓒ Ⓓ
19. Ⓐ Ⓑ Ⓒ Ⓓ
20. Ⓐ Ⓑ Ⓒ Ⓓ
21. Ⓐ Ⓑ Ⓒ Ⓓ
22. Ⓐ Ⓑ Ⓒ Ⓓ
23. Ⓐ Ⓑ Ⓒ Ⓓ
24. Ⓐ Ⓑ Ⓒ Ⓓ
25. Ⓐ Ⓑ Ⓒ Ⓓ
26. Ⓐ Ⓑ Ⓒ Ⓓ

27. Ⓐ Ⓑ Ⓒ Ⓓ
28. Ⓐ Ⓑ Ⓒ Ⓓ
29. Ⓐ Ⓑ Ⓒ Ⓓ
30. Ⓐ Ⓑ Ⓒ Ⓓ
31. Ⓐ Ⓑ Ⓒ Ⓓ
32. Ⓐ Ⓑ Ⓒ Ⓓ
33. Ⓐ Ⓑ Ⓒ Ⓓ
34. Ⓐ Ⓑ Ⓒ Ⓓ
35. Ⓐ Ⓑ Ⓒ Ⓓ
36. Ⓐ Ⓑ Ⓒ Ⓓ
37. Ⓐ Ⓑ Ⓒ Ⓓ
38. Ⓐ Ⓑ Ⓒ Ⓓ
39. Ⓐ Ⓑ Ⓒ Ⓓ

40. Ⓐ Ⓑ Ⓒ Ⓓ
41. Ⓐ Ⓑ Ⓒ Ⓓ
42. Ⓐ Ⓑ Ⓒ Ⓓ
43. Ⓐ Ⓑ Ⓒ Ⓓ
44. Ⓐ Ⓑ Ⓒ Ⓓ
45. Ⓐ Ⓑ Ⓒ Ⓓ
46. Ⓐ Ⓑ Ⓒ Ⓓ
47. Ⓐ Ⓑ Ⓒ Ⓓ
48. Ⓐ Ⓑ Ⓒ Ⓓ
49. Ⓐ Ⓑ Ⓒ Ⓓ
50. Ⓐ Ⓑ Ⓒ Ⓓ

Practice Test 2

SECTION I: MULTIPLE-CHOICE

Time: 90 minutes
50 questions

> **DIRECTIONS:** Each of the questions or incomplete statements below is followed by four suggested answers or completions. Select the one (or two where indicated) that is best in each case. You have 90 minutes to complete this portion of the test. You may use a calculator and the information sheets provided in the appendix.

1. A block ($m = 1.5$ kg) is pushed along a frictionless surface for a distance of 2.5 meters, as shown above. How much work has been done if a force of 10 newtons makes an angle of 60 degrees with the horizontal?

(A) Zero
(B) 12.5 J
(C) 21.6 J
(D) 25 J

2. What is the instantaneous power due to the gravitational force acting on a 3-kilogram projectile the instant the projectile is traveling with a velocity of 10 meters per second at an angle of 30 degrees above the horizontal?

(A) 300 W

(B) 150 W

(C) −150 W

(D) −300 W

3. A hockey puck of an unknown mass is sliding along ice that can be considered frictionless with a velocity of 10 meters per second. The puck then crosses over onto a rough floor that has a coefficient of kinetic friction equal to 0.2. How far will the puck travel before friction stops it?

(A) 2.5 m

(B) 5 m

(C) 25 m

(D) Depends on the mass

4. A dart is placed onto a spring. The spring is stretched a distance x. By what factor must the spring's elongation be changed so that the maximum kinetic energy given to the dart is doubled?

(A) 1/2

(B) 2

(C) 4

(D) $\sqrt{2}$

QUESTIONS 5 AND 6 ARE BASED ON THE FOLLOWING INFORMATION:

A 10-kilogram projectile is launched at a 60° angle to the ground, with a velocity of 200 m/s. Neglect air resistance. The ground is level.

5. Compare this projectile with a 5-kilogram projectile launched under the same conditions but at a 30° angle. The 5-kilogram projectile will

(A) go higher up and farther along the ground

(B) go equally high and equally far along the ground

(C) neither go as high nor as far along the ground

(D) not go as high but go equally far along the ground

6. As the launch angle is lowered to 45°, the maximum horizontal distance traveled by the projectile will

(A) decrease only

(B) increase only

(C) increase and then decrease

(D) decrease and then increase

7. An object of mass m rests on top of a spring that has been compressed by x meters. The force constant for this spring is k. The mass is not attached to the spring and will shoot upward when the spring is uncompressed. When released, how high will the mass rise?

    (A) $mg - kx$
    (B) kx^2/mg
    (C) $(kx^2/2mg) - x$
    (D) $(k/m)^{\frac{1}{2}} x$

8. Which of the following is the best method for finding a spring's force constant k?

    (A) Hanging a known mass on the spring and dividing the weight by the length of spring
    (B) Hanging several known masses on the spring, taking the average value of the mass, and dividing by the average length of the spring
    (C) Hanging several known masses on the spring and finding the area under the curve after plotting force versus extension
    (D) Hanging several known masses on the spring and finding the slope of the graph after plotting force versus extension

9. How much of a braking force is applied to a 2,500-kilogram car on the Moon ($g = 1.6$ m/s^2) that has an initial velocity of 30 meters per second if the car is brought to a stop in 15 seconds?

    (A) 5,000 N
    (B) 6,000 N
    (C) 8,000 N
    (D) 25,000 N

10. A 1-kilogram object is moving to the right with a velocity of 6 meters per second. It collides with and sticks to a 2-kilogram mass, which is also moving to the right, with a velocity of 3 meters per second. What happens to the total kinetic energy during this collision?

    (A) The kinetic energy is conserved because the collision is elastic.
    (B) The kinetic energy is conserved even though the collision is not elastic.
    (C) Some kinetic energy is lost during the collision even though total momentum is conserved.
    (D) Some kinetic energy is lost during the collision because of the elastic nature of the collision.

11. A ball with a mass of 0.2 kilogram strikes a wall with a velocity of 3 meters per second. It bounces straight back with a velocity of 1 meter per second. What was the magnitude of the impulse delivered to this ball?

(A) 0.2 kg · m/s

(B) 0.4 kg · m/s

(C) 0.6 kg · m/s

(D) 0.8 kg · m/s

12. Which of the following is an equivalent expression for the maximum velocity attained by a mass m oscillating horizontally along a frictionless surface? The mass is attached to a spring with a force constant k and has an amplitude of A.

(A) Ak/m

(B) $A(k/m)^{1/2}$

(C) mg/kA

(D) $A2k/m$

13. A 0.5-kilogram mass is attached to a spring with a force constant of 50 newtons per meter. What is the total energy stored in the mass-spring system if the mass travels a distance of 8 cm in one cycle?

(A) 0.5 J

(B) 0.01 J

(C) 0.04 J

(D) 0.08 J

14. Which of the following expressions is equivalent to the magnitude of the escape velocity in terms of the magnitude of the orbital velocity v for a spacecraft?

(A) $2v$

(B) v

(C) $4v$

(D) $\sqrt{2}v$

15. Which of the following graphs correctly shows the relationship between magnitude of gravitational force and distance between two masses?

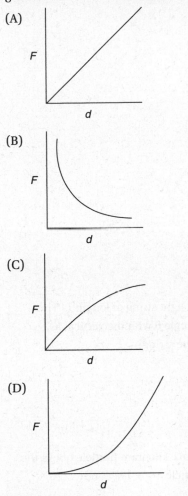

(A)

F

d

(B)

F

d

(C)

F

d

(D)

F

d

16. Standing wave nodes in a string occur every 20 cm. If the velocity of the incident waves is equal to 5 m/s and the length of the string is 80 cm, what is the frequency of the waves equal to?

(A) 25 Hz
(B) 6.25 Hz
(C) 12.5 Hz
(D) 3.125 Hz

17. Is it possible to make two wires of equal electrical resistance if one is made out of a much more conductive material?

(A) No. The more conductive material will always have a lower resistance.
(B) No. The less conductive material will always require a greater voltage to produce an equal current.
(C) Yes. The less conductive wire must be made with a larger cross-sectional area.
(D) Yes. The more conductive wire must be shorter.

18. A projectile is launched at a 30° angle to the ground with a velocity of 200 m/s. What is its speed at its maximum height?

(A) 9.8 m/s

(B) 100 m/s

(C) 173 m/s

(D) 200 m/s

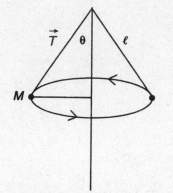

19. A conical pendulum consists of a mass m attached to a light string of length L. The mass swings around in a horizontal circle, making an angle θ with the vertical as shown above. What is the magnitude of tension, T, in the string?

(A) $mg/\cos\theta$

(B) $mg\cos\theta$

(C) $mg/\sin\theta$

(D) $mg\sin\theta$

20. The magnitude of the one-dimensional momentum of a 2-kilogram particle obeys the relationship $p = 2t + 3$. What was the velocity of the particle at $t = 1$ second?

(A) 5 m/s

(B) 2 m/s

(C) 1 m/s

(D) 2.5 m/s

21. An object is experiencing a nonzero net force. Which of the following statements is most accurate?

(A) The linear and angular momentums of the object are both definitely changing.

(B) Although the linear momentum of the object is definitely changing, the angular momentum may not be.

(C) Although the angular momentum of the object is definitely changing, the linear momentum may not be.

(D) Neither the linear momentum nor the angular momentum is definitely changing.

22. A car with a 500-newton driver goes over a hill that has a radius of 50 meters as shown above. The velocity of the car is 20 meters per second. What are the approximate force and the direction that the car exerts on the driver?

(A) 900 N, up
(B) 400 N, down
(C) 100 N, up
(D) 500 N, up

23. What is the net torque acting on the pivot supporting a stationary 10-kilogram beam 2 meters long as shown above?

(A) 198 N · m
(B) −198 N · m
(C) −102 N · m
(D) 102 N · m

24. A trumpet player notices a string on a harp is beginning to vibrate when he plays a certain note on his trumpet. What is this an example of?

(A) Constructive interference
(B) Destructive interference
(C) Refraction
(D) Resonance

25. What is the ratio of the equivalent resistance of 2 resistors R in series to the same 2 resistors in parallel?

(A) 0.5
(B) 1
(C) 2
(D) 4

26. A 12-volt battery is advertised as a "40 amp hour" battery. The manufacturer is expressing that the battery has a capacity of

(A) 144,000 C

(B) 144,000 J

(C) 1,728,000 J

(D) 1,728,000 W

27. Compare the force experienced by an electron in the two situations shown above. In the situation on the left, the electron is a distance D from a positive charge Q. In the situation on the right, the electron is 3 times farther ($3D$) from 3 times as much positive charge ($3Q$) as in the scenario on the left. The force of electrical attraction in the scenario on the right will

(A) remain the same as the situation on the left

(B) be 3 times stronger than the situation on the left

(C) be 3 times weaker than the situation on the left

(D) be 9 times weaker than the situation on the left

28. An old record player could bring a disk up to its 45 RPM speed in less than a second. If the same size disk can also be brought up to a speed of 75 RPM in about the same amount of time on another player, compare the two torques.

(A) The torques would be the same as the moment of inertia of the two disks are the same.

(B) The torques would be the same because of the conservation of angular momentum.

(C) The torque would be larger in the second case as it requires a greater angular acceleration.

(D) The torque would be larger in the second case as it entails both a larger force and a larger lever arm.

Two small, identical metal spheres are projected at the same time from the same height by two identical spring guns. Each gun provides the same push on its sphere. However, one sphere is projected vertically upward while the other sphere is projected horizontally. The speed of each projectile as it emerges from the gun is the same. Frictional losses are negligible.

29. How does the speed of the vertically launched sphere compare to the speed of the horizontally launched sphere as they each hit the floor?

    (A) It is the same.
    (B) It is twice as great.
    (C) It is greater but not necessarily twice as great.
    (D) It is less.

30. How does the time required for the vertically projected sphere to hit the floor compare with that for the horizontally projected sphere?

    (A) It is the same.
    (B) It is twice as great.
    (C) It is greater but not necessarily twice as great.
    (D) It is less.

31. How does the work done by gravity to the vertically launched sphere compare to the work done by gravity to the horizontally launched sphere?

    (A) It is the same.
    (B) It is twice as great.
    (C) It is greater but not necessarily twice as great.
    (D) It is less.

QUESTIONS 32 AND 33 REFER TO THE FOLLOWING INFORMATION:

A racing car is speeding around a flat, circular track whose radius is 250 m. The car's speed is a constant 50 m/s, and the car has a mass of 2,000 kg.

32. The centripetal force necessary to keep the car in its circular path is provided by

    (A) the engine
    (B) the brakes
    (C) friction
    (D) the steering wheel

33. The magnitude of the car's centripetal force is equal to

    (A) 10 N
    (B) 400 N
    (C) 4,000 N
    (D) 20,000 N

Two small masses, X and Y, are d meters apart. The mass of X is 4 times as great as the mass of Y. Mass X attracts mass Y with a gravitational force of 16 N.

34. The force with which Y attracts X is equal to

 (A) 4 N
 (B) 16 N
 (C) 32 N
 (D) 64 N

35. If the distance between X and Y is doubled, then X will attract Y with a force of

 (A) 4 N
 (B) 8 N
 (C) 16 N
 (D) 32 N

36. A projectile is launched at an angle such that it undergoes projectile motion. In the absence of any air resistance, which of the following statements is correct?

 (A) The horizontal velocity increases during the flight.
 (B) The horizontal velocity remains constant during the flight.
 (C) The horizontal velocity decreases during the flight.
 (D) The vertical acceleration changes during the flight.

37. Which of the following charges is NOT possible for an isolated particle?

 (A) 4.8×10^{-19} C
 (B) 4.8×10^{-18} C
 (C) 4.8×10^{-20} C
 (D) -4.8×10^{-19} C

38. The sound of a moving siren from the west of you that is traveling toward you is transmitted through the air to your ear and is

 (A) vibrating in a north-south direction
 (B) vibrating in a west-east direction
 (C) vibrating up and down
 (D) vibrating in a corkscrew fashion

39. The sound of a moving siren from the west of you that is traveling toward you is transmitted through the air to your ear and is

 (A) moving faster than the usual speed of sound
 (B) heard at a lower frequency than usual
 (C) heard at a higher frequency than usual
 (D) heard at a lower amplitude than usual

40. A woman standing on a scale in an elevator notices that the scale reads her true weight. From this, she may conclude that

(A) the elevator must be at rest
(B) the elevator must be accelerating precisely at 9.8 m/s^2
(C) the elevator must be moving upward
(D) the elevator must not be accelerating

41. A planet has half the mass of Earth and half the radius. Compared with the acceleration due to gravity at the surface of Earth, the acceleration due to gravity at the surface of this planet is

(A) the same
(B) halved
(C) doubled
(D) quadrupled

42. Complete the analogy.

Force is to torque as _____ is to moment of inertia.

(A) momentum
(B) angular momentum
(C) angular acceleration
(D) mass

43. While standing on a moving train, you are suddenly thrown forward as the train stops. This effect can be best explained using Newton's law of

(A) inertia
(B) acceleration
(C) action and reaction
(D) universal gravity

44. For a falling object to reach terminal velocity (a constant downward speed), the magnitude of the force due to air friction must be

(A) negligible
(B) less than mg
(C) equal to mg
(D) greater than mg

45. An object is in free fall for T seconds. Compare its change in velocity during the first $T/2$ seconds with its change in velocity during the final $T/2$ seconds.

(A) They are the same.
(B) They are larger during the first interval.
(C) They are smaller during the first interval by a factor of ½.
(D) They are less than ½ as small during the first interval.

46. Which of the following must be zero if an object is spinning at a constant angular velocity? Select two correct answers.

(A) Net torque
(B) Moment of inertia
(C) Angular momentum
(D) Angular acceleration

47. Which of the following could be used to calculate the time in flight for a horizontally launched projectile on Earth? Ignore friction. Select two correct answers.

(A) The launch speed
(B) The final vertical velocity
(C) The final horizontal velocity
(D) The initial height

48. As an ice skater draws her arms inward during a spin, which of the following remain constant? Select two correct answers.

(A) Her moment of inertia
(B) Her angular momentum
(C) Her angular velocity
(D) The location on the surface of the ice of her center of mass (ignoring any positional changes)

49. A total of three forces, each with a magnitude of 5 N, are exerted on an object with a mass of 15 kg. Which of the following statements are true? Select two correct answers.

(A) The object may be accelerating at 1 m/s^2.
(B) The object might not be accelerating.
(C) The object might be accelerating at 2 m/s^2.
(D) The sum of the components of the object's acceleration must add up to the magnitude of its acceleration $(a_x + a_y = |\vec{a}|)$.

50. In the circuit above, with the switch closed, what will voltmeter V read? (A is an ammeter.) Select two answers.

(A) The voltage drop across R_1
(B) The voltage drop across ammeter A
(C) The voltage drop across R_2
(D) Zero

SECTION II: FREE-RESPONSE

Time: 90 minutes

5 questions

> **DIRECTIONS:** You have 90 minutes to complete this portion of the test. You may use a calculator and the information sheets provided in the appendix.

1. (7 points; ~13 minutes) An elevator ride consists of the following portions.

   (a) Initially, the elevator is on the ground floor and quickly speeds up in the upward direction on its way to the top floor.

   (b) Most of the trip upward is at a constant velocity.

   (c) Suddenly, near the top floor, the elevator comes to a stop just in time for the top floor.

   (d) Now the button for the first floor is pressed, and the elevator speeds downward suddenly.

   (e) Most of the trip downward is spent at constant velocity.

   (f) Suddenly, at the very bottom, the elevator comes to a stop.

   (a) Draw a qualitative free-body diagram (showing the relative vector sizes) for each portion of the elevator ride as experienced by a passenger. Next to each, describe how the person would feel (normal, heavier than usual, lighter than usual, etc.).

   (b) Explain the difference among the passenger's mass, gravitational weight, and apparent weight throughout the trip. When (if ever) are these the same, and when (if ever) are they different? Which quantities change during the trip, and which ones do not?

   (c) Describe what would happen if, on the way up (during step (b)), the cables were to break. Ignore any frictional effect between the elevator itself and the elevator shaft. Be sure to include a discussion of the relative motion of the elevator car and the passenger. Your answer should include a discussion or sketch of the height above ground, the velocity, and the acceleration experienced by the passenger from the time of the break until the end of the motion.

   (d) Sketch the expected behavior of the person's height as a function of time for situation (c) above. Also sketch the person's V_y as a function of time as well.

2. (12 points; ~25 minutes) A group of students have been asked to determine the spring constant k of the following setup. They have a known mass and a very smooth tabletop. The only additional equipment they have is metersticks and rulers.

(a) Describe an experimental procedure to determine the spring constant k. You may wish to label the diagram above further to help in your description. Indicate what measurements you would take and how you would take them. Include enough detail so that another student could carry out your procedure.

(b) Determine your independent and dependent variables in order to create a linear plot to determine k. Sketch out the expected graph. Explain your reasoning.

(c) What common sources of error in your procedure might happen during this investigation? What steps could you take to minimize these errors?

(d) A group of students collected the following data; x is for the spring and D is along the ground. The students found a discontinuity in the behavior of the graph. Graph the data and determine a value for k. Offer a possible explanation for the discontinuity.

x (cm)	D (m)
1	0.21
2	0.41
3	0.59
4	0.8
5	0.99
6	1.19
7	1.22
8	1.28
9	1.34
10	1.46

3. (12 points; ~25 minutes) Two cars are separated by 25 meters. Both are initially at rest. Then the car in front begins to accelerate uniformly at x meters per second squared. The second car, which is behind, begins to accelerate uniformly at y meters per second squared.

   (a) If the second car must catch up with the car in front within 10 seconds, what type of relationship must exist between y and x? Is there a maximum or minimum difference between the two accelerations? Explain using words and equations.

   (b) Given that the conditions in question (a) are being met (that is, the second car will catch up to the first within 10 seconds), is there any instant in time that the cars will have the same velocity? Will there be any interval of time over which both cars will have the same average velocity? Support your answers with graphs.

   (c) Given that the conditions in question (a) are being met (i.e., the second car will catch up to the first within 10 seconds) and that both cars are continuing to drive with constant velocity for an additional 10 seconds at their instantaneous speeds at the moment the second car has caught up with the first, which car will be ahead and by how much by the end of the trip? Answer this question qualitatively and quantitatively. Specifically determine the minimum and maximum differences in displacement if they can be found.

(d) If both cars are on track to meet the conditions in question (a) (i.e., the second car will catch up to the first within 10 seconds) but both cars are stopped after t raveling 50 meters, which one will have expended the most energy? Assume both cars are identical in mass and fuel efficiency. Choose one of the following options, and justify your choice:

_____ The first car expended more energy.

_____ Both cars expended equal energy.

_____ The second car expended more energy.

_____ Which car expended more energy requires more knowledge of exactly when the second car catches up to the first car.

4. (7 points; ~13 minutes) A long-distance, unmanned probe is sent to another planet. To keep it oriented correctly, the probe is given a lot of spin during the early part of the trip. However, when the probe is getting close to its destination, the spin is no longer desired. Two very long cables with mass at the end of each are extended while the probe is still in deep space. Once extended, the cables are released and the probe begins its landing sequence.

(a) Explain qualitatively why the spinning will help keep the probe oriented correctly. Describe the underlying physics concepts using equations and diagrams as needed.

(b) Explain what is accomplished by the action of extending and releasing the cables near the end of the probe's flight. Describe the underlying physics concepts using equations and diagrams as needed. Do the masses need to be a significant portion of the probe's mass for this procedure to be effective?

(c) What happens to all of the probe's linear momentum when it lands on its destination planet? Describe qualitatively in terms of physics principles what happens during the landing.

5. (7 points; ~13 minutes)

(a) A fellow student argues that you cannot use the formula $\frac{1}{2}gt^2$ to find the height of a very tall cliff (200 meters or so) when dropping a stone from the top and counting the seconds until impact at the bottom. He says that because of Newton's law of universal gravity, the acceleration is not constant during the fall. Using equations and calculations, both support and refute his argument. Would his argument have had greater, lesser, or the same power to persuade you if he was talking about a cliff on the Moon? Answer in a clear, coherent paragraph. Use equations to support your reasoning.

(b) Two masses, M and $9M$, are separated by a distance of d. At what distance and in what direction from the smaller mass should a third mass be placed such that the net gravitational force on this third mass is zero?

1.	**B**	14.	**D**	27.	**C**	40.	**D**
2.	**C**	15.	**B**	28.	**C**	41.	**C**
3.	**C**	16.	**C**	29.	**A**	42.	**D**
4.	**D**	17.	**C**	30.	**C**	43.	**A**
5.	**D**	18.	**C**	31.	**A**	44.	**C**
6.	**B**	19.	**A**	32.	**C**	45.	**A**
7.	**C**	20.	**D**	33.	**D**	46.	**A, D**
8.	**D**	21.	**B**	34.	**B**	47.	**B, D**
9.	**A**	22.	**C**	35.	**A**	48.	**B, D**
10.	**C**	23.	**D**	36.	**B**	49.	**A, B**
11.	**D**	24.	**D**	37.	**C**	50.	**A, C**
12.	**B**	25.	**D**	38.	**B**		
13.	**B**	26.	**C**	39.	**C**		

ANSWERS EXPLAINED

Section I: Multiple-Choice

1. **(B)** Work $= fd\cos 60° = (10 \text{ N})(2.5)(0.5) = 12.5 \text{ J}$

2. **(C)**

$$\text{Power} = \text{work}/\text{time} = (fd\cos 120°)/T = (mg\cos 120°)(D/T)$$
$$= (3)(10)(-0.5)(10 \text{ m/s}) = -150 \text{ W}$$

3. **(C)**

$$W = \Delta KE$$
$$\mu ND\cos 180° = 0 - \frac{1}{2}mv^2$$
$$(\mu mgD)(-1) = -\frac{1}{2}mv^2$$
$$D = v^2/2\mu g = 100/(2 \times 0.2 \times 10) = 25 \text{ m}$$

4. **(D)** Conservation of energy:

$$KE = EPE = \frac{1}{2}kx^2$$

Since we need x^2 to double, increase x by $(2)^{1/2} = \sqrt{2}$.

5. **(D)** The 30° and 60° angles will have equal ranges but with the roles of V_x and V_y reversed. Therefore, the projectile launched at a 30° angle will not go as high. Note that mass does not enter into projectile motion problems.

6. **(B)** 45° gives the maximum range for a projectile as it is splitting the initial velocity evenly between vertical (giving you time in flight) and horizontal (giving you speed downrange). Angles above or below will have shorter ranges. Approaching the 45° angle will increase the range:

$$\text{Range} = \left(\frac{v_0^2}{g}\right)\sin 2\theta$$

7. **(C)**

$$\frac{1}{2}kx^2 = mgh$$

$h = kx^2/2mg$ above the starting point. Since the release point is a distance x above the starting point, we must subtract x from the answer.

8. **(D)** $F = kx$. So when graphing f versus x, k will be the slope. The extension of the spring is x when F is the force applied to the spring.

9. **(A)**

$$\vec{F}\Delta t = \Delta \vec{p}$$
$$(\vec{F})(15 \text{ s}) = 0 - (2{,}500)(30) = -75{,}000 \text{ kg} \cdot \text{m/s}$$
$$\vec{F} = -5{,}000 \text{ N}$$

The negative sign indicates an opposing force. Note that weight is not needed in this calculation, so the value of g is irrelevant.

10. **(C)** Momentum is always conserved. However, kinetic energy is lost unless the collision is elastic, in which case the kinetic energy is also conserved:

$$\vec{p}_i = 1(+6) + 2(+3) = 12 \text{ kg m/s} = \vec{p}_f = 3\vec{v}_f$$
$$\vec{v}_f = 4 \text{ m/s}$$

Initial *KE*:

$$\left(\frac{1}{2}\right)(1)(6)^2 + \left(\frac{1}{2}\right)(2)(3)^2 = 27 \text{ J}$$

Final *KE*:

$$\left(\frac{1}{2}\right)(3)(4)^2 = 24 \text{ J}$$

3 joules are lost.

11. **(D)** Impulse $= \Delta p = p_f - p_i = 0.2(-1) - (0.2)(3) = -0.8$ kg $\cdot$ m/s

12. **(B)**

$$\frac{1}{2}kA^2 = \text{total energy when } x = 4 \, (\text{all } PE, \text{no } KE)$$

$$\frac{1}{2}mv_{\text{max}}^2 = \text{total energy when } x = 0 \, (\text{all } KE, \text{no } PE)$$

Conservation of energy:

$$\frac{1}{2}kA^2 = \frac{1}{2}mv_{\text{max}}^2$$

$$v_{\text{max}} = (k/m)^{\frac{1}{2}}A$$

13. **(B)** In one cycle, the mass travels 4 amplitudes:

$$A = 0.02 \text{ m}$$

$$\text{Energy} = \frac{1}{2}kA^2 = \frac{1}{2}(50)(0.02)^2 = 0.01 \text{ J}$$

14. **(D)** Orbital velocity:

$$F_g = \text{centripetal force} = mv^2/r$$

$$GMm/r^2 = mv^2/r$$

$$v = (GM/r)^{\frac{1}{2}}$$

Escape velocity:

$$E_{\text{total}} > 0$$

$$KE + PE > 0$$

$$\frac{1}{2}mv^2 - GMm/r > 0 \, (\text{using universal gravitational } PE)$$

$$v > (2GM/r)^{\frac{1}{2}}$$

15. **(B)** The gravitational force between two masses has a $1/r^2$ relationship (where r is the distance from center to center). Graph B shows the correct inverse relationship.

16. **(C)** Since node to node is half a wavelength:

$$\lambda = 0.4 \text{ m}$$

$$f = v/\lambda = 5/0.4 = 12.5 \text{ Hz}$$

17. **(C)** Resistivity is only part of the resistance of a wire:

$$R = \rho L / A$$

Therefore, a less conductive material (lower ρ) can be made equal in resistance to a more conductive material by making the less conductive material shorter and/or fatter (lower L or larger A).

18. **(C)** At maximum height, all speed is from V_x:

$$V_x = 200\cos 30° = 173 \text{ m/s}$$

19. **(A)** Vertical forces must cancel:

$$T\cos\theta - mg = 0$$
$$T = mg / \cos\theta$$

20. **(D)** $v = p/m = (2 \times 1 + 3)/2 = 2.5$ m/s

21. **(B)** Since $\vec{F}\Delta t = \Delta\vec{p}$, a net force will change the linear momentum. However, torque also involves the lever arm ($\tau = RF\sin\theta$). So despite having a nonzero net force, the net torque might still be zero.

22. **(C)** Circular motion:

$$F_{net} = mv^2 / r$$
$$mg - N = mv^2 / r$$

N is the force the car exerts on the driver:

$$mg - mv^2/r = 500 - 50(20^2)/50 = 500 - 400 = 100 \text{ N}$$

Note that the correct answer can be found by simply knowing that the net force must be down and that the car must be pushing upward on the passenger.

23. **(D)** Take the center of mass to be 1 meter from the pivot:

$$\text{Net torque} = -mg(1)\sin 90° + 200(2)\sin 30°$$
$$= 98 + 200 = 102 \text{ N·m}$$

24. **(D)** Resonance is the sympathetic vibration of an object when impacted by a wave with the same frequency as its own fundamental frequency of oscillation.

25. **(D)**

$$\text{Resistors in series} = 2R$$
$$\text{Resistors in parallel} = R/2$$
$$\text{Difference} = 1.5R$$
$$\text{Ratio} = 2R/(R/2) = 4$$

26. **(C)** Batteries store energy, which is measured in joules:

$$\text{Energy} = \text{power} \cdot \text{time} = I \cdot V \cdot \text{time} = (40\,\text{A})(12\,\text{V})(3{,}600\,\text{s})$$
$$= 1{,}728{,}000\,\text{J}$$

27. **(C)** Coulomb's law is kQ_1Q_2/r^2. Tripling Q_1 triples the force. Tripling r makes the force 9 times weaker. Net effect:

$$3 \times 1/9 = 1/3$$

28. **(C)**

$$\text{Torque} = I\alpha$$
$$\alpha = \text{angular acceleration} = \Delta\omega/\Delta t$$

Our only data provided are that the angular acceleration must be larger for the 75 RPM record and that the moment of inertia has not changed (same disk). Therefore, the torque supplied must be larger.

29. **(A)** Remember conservation of energy. Since both start at the same height (same potential energy) with the same kinetic energy, they will both hit the ground with the same joules of energy (all kinetic energy). The same kinetic energy means the same speed since their masses are the same. Note that this does not imply that both components of velocity are the same. They are not. This implies that only the magnitudes of the final velocity vectors are the same.

30. **(C)** The vertically projected sphere will spend much more time in the air as it goes much higher. Without knowing the exact speed of the launch, It Is not possible to say by what factor the time in flight is extended.

31. **(A)** The change in gravitational potential energy is the same for both. Therefore, the work done by gravity is the same.

32. **(C)** Friction is the only force that is directed inward toward the center of the circle. By definition, all centripetal forces must be directed inward. To confirm this, imagine the path the car would take on a frictionless stretch of track.

33. **(D)** $mv^2/r = (2,000 \times 50^2)/250 = 20,000$ N

34. **(B)** Remember Newton's third law. The forces are equal and opposite.

35. **(A)** Universal gravity is a $1/r^2$ law. Doubling r makes the force 4 times weaker.

36. **(B)** No horizontal acceleration means no change in horizontal velocity.

37. **(C)** Charge is quantized (comes in integer multiples only) in units of $\pm 1.6 \times 10^{-19}$ C.

38. **(B)** Sound is a longitudinal wave.

39. **(C)** The Doppler shift of waves that occurs when the source and receiver are moving toward each other causes a shorter wavelength and hence a higher frequency.

40. **(D)** The elevator must be traveling at constant velocity to insure the normal force (reading on the scale) is the same as her true weight. However, this velocity value can be any number: positive, zero, or negative.

41. **(C)** Remember that $g = Gm/r^2$. Half the mass means half the g. Half the r means $4 \times g$. Collect the changes:

$$\frac{1}{2} \times 4 = 2$$

42. **(D)** Forces cause mass to accelerate linearly. Torques cause moments of inertia to accelerate angularly.

43. **(A)** Inertia is the tendency of an object to continue its motion in the absence of other forces. As the train stops, the passengers continue forward until a force is brought to bear directly on them.

44. **(C)** Constant velocity means no acceleration. No acceleration means no net force. The friction opposing weight must be equal in magnitude to that weight.

45. **(A)** Constant acceleration means precisely that. For the same time interval, the change in velocity will be the same.

46. **(A) and (D)** Constant angular velocity means zero angular acceleration, which means no net torque.

47. **(B) and (D)** Since the vertical velocity is zero, only height is needed to determine the drop time:

$$H = \frac{1}{2}gt^2$$

Knowing the final vertical velocity would also allow one to determine the time since:

$$V_{yf} = 0 - gt$$

48. **(B) and (D)** The skater's center of mass must remain above her skates. Since no external torques are involved in drawing her arms inward, her momentum is conserved. Note that her moment of inertia decreases and her angular velocity increases.

49. **(A) and (B)** If the 3 forces are in the same direction, F_{net} = 15 N, which is the maximum possible force. In this situation, the acceleration is 1 m/s². If the 3 forces are 120° apart from each other in direction, they would add up to zero and produce no acceleration.

50. **(A) and (C)** Both R_1 and R_2 are connected in parallel to the same two points that the voltmeter is measuring the potential difference between. Therefore, the voltmeter reading is that same voltage drop for both R_1 and R_2. Note that an ideal ammeter has no resistance and thus experiences no voltage drop itself.

Section II: Free-Response

1. (7 points)

   (a) (3 points)

      a. (0.5 points) $\vec{N}$ Upward acceleration, feel heavy

$m\vec{g}$

      b. (0.5 points) $\vec{N}$ No acceleration, feel normal

$m\vec{g}$

c. (0.5 points) $\vec{N}$ Downward acceleration, feel light

$m\vec{g}$

d. (0.5 points) $\vec{N}$ Downward acceleration, feel light

$m\vec{g}$

e. (0.5 points) $\vec{N}$ No acceleration, feel normal

$m\vec{g}$

f. (0.5 points) $\vec{N}$ Upward acceleration, feel heavy

$m\vec{g}$

(b) (1 point) A person's mass is a measure of his or her inertia. This value does not change due to acceleration or changes in location. Gravitational weight is a force due to the interaction of the person's mass and the planet on which he or she is standing (including the distance between their centers). Although technically this value is slightly smaller as you get higher above sea level, the differences within a building on Earth are negligible. Apparent weight is the contact force your body experiences, which gives you your subjective experience of "weight." In this case, the normal force and the changing values of the normal force explain the changes the person would experience on the elevator ride.

(c) (1 point) The elevator and passenger would experience free fall. The only force would be the downward mg, and the normal force would be zero. Hence, the person would feel weightless. Since the car was on the way up when the cables broke, both passenger and elevator would maintain the same relative velocity to each other as both continued upward, slowed down, and then reversed direction and continued to speed up while falling. The entire time, the passenger would feel weightless.

(d) (2 points; 1 point for each graph)

2. (12 points)

(a) (2 points) For various measured compressions of the spring (x), measure horizontal range for the mass (R). Range should be measured along the floor from beneath the edge of the table to where the mass first hits the ground. Multiple trials for each compression x should be taken so that the average range of values can be determined.

(b) (4 points) The independent variable is the one the experimenter controls and manipulates directly. In this case, the independent variable is the compression x. The dependent variable is the one measured as a result of changes in the independent variable. In this case, the dependent variable is the range. Independent variables are graphed on the horizontal axis. (1 point)

Theoretical prediction:

$$\frac{1}{2}kx^2 = \frac{1}{2}mv^2$$

$$v = (k/m)^{\frac{1}{2}}x \qquad \text{(1 point)}$$

This velocity is the horizontal projectile's velocity. The time in flight is found from the height of the table:

$$H = \frac{1}{2}gt^2$$

$$1.6\text{ m} = 4.9t^2$$

$$t = 0.57\text{ s}$$

$$R = vt = (k/m)^{1/2}xt = .57(k/m)^{1/2}x \qquad \text{(1 point)}$$

Using the fixed values for m and t:

$$R = 0.806k^{1/2}x \qquad \text{(1 point)}$$

(c) (1 point) The major source of error is friction of the tabletop between the end of the spring and the edge of the table. Ensuring that this distance is small and that the surfaces involved are smooth will minimize this error.

(d) (5 points; 2 points for labels with scaling, 1 point for best fit, 1 point for a correct explanation of discontinuity, 1 point for solving for k)

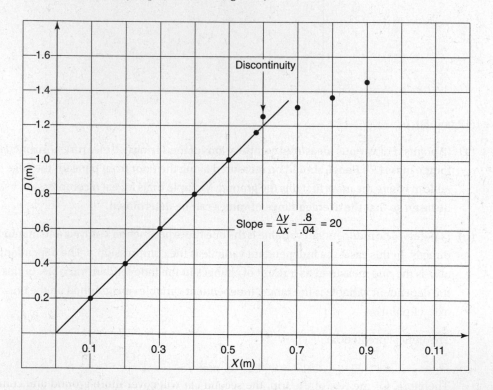

The discontinuity is probably caused by exceeding the limit of elasticity for this spring. Hooke's law assumes that the material is perfectly elastic and resumes its shape after being stretched or compressed:

$$\text{Slope} = 20 = 0.806k^{1/2} \text{ (from part (b))}$$

Solving for k:

$$k = 616\text{ N/m}$$

3. (12 points)

(a) (4 points) When the cars meet, the position of each car, relative to a common origin, must be the same. (1 point) Since each car is starting from rest and accelerating uniformly:

$$d_1 = d_2$$
$$\frac{1}{2}xt^2 + 25 = \frac{1}{2}yt^2$$
$$y - x = 50/t^2 \qquad \text{(1 point)}$$

Clearly, the second car must have a larger acceleration ($y > x$). The difference between the two accelerations must increase as time to catch up (t) gets smaller. So there is no upper limit on how much faster car y must accelerate than car x to catch up within 10 seconds. (1 point) There is, however, a lower limit on the difference. This can be found by examining the case when the second car takes the full 10 seconds:

$$y - x = 50/t^2 = 50/100 = 0.5$$

Thus, we see that for the second car to catch up with the lead car within the first 10 seconds, the second car must accelerate at least 0.5 m/s^2 faster than the first car. (1 point)

(b) (3 points) No, they will never have either the same instantaneous speed or the same average speed. (1 point) The second car is always faster:

The second car has a slope of y on this graph, while the first car has a slope of x. Since $y > x$, one can see that neither the instantaneous nor the average slope of either of these plots is ever the same. (1 point)

(c) (3 points) The second car will be ahead as it will have the higher speed when the two meet. Therefore, for the rest of the trip, the second car will cover more ground and come out ahead. (1 point) To find out by how much, we can compare their speed when they meet:

$$v_1 = xt \qquad v_2 = yt$$

where t is the exact time (within the first 10 seconds) when they meet.

$$\Delta d = v_2(10) - v_1(10) = 10(yt - xt) = (y - x)10t$$

Substitute our algebraic expression from part (a):

$$\Delta d = 500/t \qquad \text{(1 point)}$$

Once again, there is no upper bound on the difference in distance. However, there is a lower bound that can be found by setting t equal to 10 seconds:

$$\Delta d > 50 \text{ m} \qquad \text{(1 point)}$$

(d) (2 points) __X__ The second car expended more energy. (1 point)

No matter the details, more work was done in moving the second car as it had the greater acceleration. Greater acceleration (with the same mass) means greater force. Since the displacements are the same, the greater force means more work:

$$W = Fd \qquad \text{(1 point)}$$

4. (7 points)

(a) (2 points) The spinning probe acts as a gyroscope since angular momentum is both a vector and conserved. The direction of the angular momentum requires an external torque to be changed. Therefore, barring some outside force, the probe will maintain the orientation it has when the angular velocity is given to its moment of inertia:

$$L = I\omega$$

(b) (3 points) By extending the masses outward from the probe's body, the moment of inertia of the spinning probe can be greatly increased. (1 point) Note that the masses need not be that great in order to change the moment of inertia substantially as the distance, r, from the axis of rotation can be made quite large by using a long cable and the r term is squared in the moment of inertia calculation:

$$I = mr^2 \qquad \text{(1 point)}$$

Increasing the moment of inertia serves to slow the spin rate of the probe by decreasing the angular velocity in order to conserve angular momentum:

$$L = I_{\text{small}}\omega_{\text{large}} = I_{\text{large}}\omega_{\text{small}} \qquad \text{(1 point)}$$

(c) (2 points) Linear momentum ($m\vec{v}$) of the probe-planet system must be conserved. Equal and opposite impulses ($F\Delta t$) are delivered to the probe and the planet during impact. However, because of their vastly different masses, what is a major impulse to the probe turns out to be a moderate impulse to the entire planet and has almost no perceptible effect on the planet's motion:

$$M_{\text{probe}}\Delta\vec{v}_{\text{big}} - M_{\text{big}}\Delta\vec{v}_{\text{small}} = 0$$

5. (7 points)

(a) (5 points) First, technically your fellow student is correct. As the top of the cliff is 200 meters farther from the center of Earth, the gravitational field experienced by the rock at the top of the cliff is technically weaker (1 point):

$$g_{top} = GM_{Earth}/(r + 200)^2 \quad \text{versus} \quad g_{bottom} = GM_{Earth}/(r)^2 \quad \text{(1 point)}$$

Since the radius of Earth is 6.38×10^6 m, adding a mere 200 m does not make much of a difference:

$$200/(6.38 \times 10^6) = 3 \times 10^{-5} \quad \text{(1 point)}$$

Therefore, for all practical purposes, the difference in gravitational field strength is so minimal that treating g as a constant acceleration is fairly reasonable.

On the Moon, one would expect a greater deviance since 200 meters is a greater fraction of the Moon's radius (1.7×10^6). (1 point) However, any decrease in the Moon's gravitational effects would also be negligible:

$$200/(1.7 \times 10^6) = 1 \times 10^{-4} \quad \text{(1 point)}$$

(b) (2 points) Somewhere along a line between M and $9M$ but closer to M should be a position such that the force of attraction for M will cancel the force of attraction for $9M$. Let the distance between the third mass (m) and M be x:

$$GmM/x^2 = Gm(9M)/(d - x)^2 \quad \text{(1 point)}$$

Cancel and simplify:

$$9x^2 = (d - x)^2$$

Take the square root of both sides:

$$\pm 3x = d - x$$

This leads us to two solutions:

$$x = d/4 \quad \text{(1 point)}$$

which is the expected solution between the masses and

$$x = -d/2$$

which is an unexpected solution on the outside of mass M along the same line. Note that this solution is nonphysical since the direction of the forces is not taken into consideration.

TEST ANALYSIS

PRACTICE TEST 2

Section 1: Multiple-Choice

Note that the questions requiring two answers are to be graded as completely correct (1 point) or incorrect (0 points, even if you have one of the two answers correct).

$$\text{Number correct (out of 50)} = \frac{}{\text{Multiple-Choice Score}}$$

Section 2: Free-Response

Partial credit is awarded for any correct responses within an individual free-response question.

$$\text{Question 1} = \frac{}{\text{(out of 7)}}$$

$$\text{Question 2} = \frac{}{\text{(out of 12)}}$$

$$\text{Question 3} = \frac{}{\text{(out of 12)}}$$

$$\text{Question 4} = \frac{}{\text{(out of 7)}}$$

$$\text{Question 5} = \frac{}{\text{(out of 7)}}$$

$$\text{Total 4} = \frac{}{\text{(out of 45)}}$$

Final Score

$$\frac{}{\text{Multiple-Choice Score}} + \left(1.11 \times \frac{}{\text{Free-Response Score}}\right) = \frac{}{\text{Total (out of 100)}}$$

Final Score Range*

Final Score Range	AP Score
71–100	5
56–70	4
41–55	3
26–40	2
0–25	1

*Note: The guidelines above are based on the released scores for past AP Physics 1 exams. Actual score ranges vary from year to year and are determined by the College Board each year. Thus, the ranges shown are approximate.

Appendix

Table of Information for AP Physics 1

Useful Constants

Rest mass of the proton	$m_p = 1.67 \times 10^{-27}$ kg
Rest mass of the neutron	$m_n = 1.67 \times 10^{-27}$ kg
Rest mass of the electron	$m_e = 9.11 \times 10^{-31}$ kg
Magnitude of the electron charge	$e = 1.60 \times 10^{-19}$ C
Speed of light	$c = 3 \times 10^8$ m/s
Coulomb's law constant	$k = (1/4)\pi\varepsilon_0 = 9 \times 10^9$ N $\cdot$ m^2/C^2
Acceleration due to gravity at Earth's surface	$g = 9.8$ m/s^2
Universal gravitational constant	$G = 6.67 \times 10^{-11}$ m^3/(kg $\cdot$ s^2)

Unit Symbols

Meter, m	Hertz, Hz	Watt, W	Degree Celsius, °C
Kilogram, kg	Newton, N	Coulomb, C	
Second, s	Joule, J	Volt, V	
Ampere, A		Ohm, Ω	
Kelvin, K			

Prefixes

Factor	Prefix	Symbol
10^{12}	Tera-	T
10^{9}	Giga-	G
10^{6}	Mega-	M
10^{3}	Kilo-	k
10^{-2}	Centi-	c
10^{-3}	Milli-	m
10^{-6}	Micro-	μ
10^{-9}	Nano-	n
10^{-12}	Pico-	p

Formula Sheet for AP Physics 1

Mechanics

$$v = v_0 + at$$

$$x = x_0 + v_0 t + \frac{1}{2}at^2$$

$$v^2 = v_0^2 + 2a(x - x_0)$$

$$\sum \vec{F} = \vec{F}_{net} = m\vec{a}$$

$$F_{fric} \leq \mu N$$

$$a_c = \frac{v^2}{r}$$

$$\tau = rF\sin\theta$$

$$\vec{p} = m\vec{v}$$

$$\vec{F}\Delta t = \Delta \vec{p}$$

$$K = \frac{1}{2}mv^2$$

$$\Delta U_g = mg\Delta y$$

$$\Delta E = W = Fd\cos\theta$$

$$P_{avg} = \frac{\Delta E}{\Delta t}$$

$$\theta = \theta_0 + \omega_0 t + \frac{1}{2}\alpha t^2$$

$$\omega = \omega_0 + \alpha t$$

$$x = A\cos(2\pi ft)$$

$$\sum \vec{\tau} = \vec{\tau}_{net} = I\vec{\alpha}$$

$$T = rF\sin\theta$$

$$L = I\omega$$

$$\Delta L = \tau \Delta t$$

$$K = \frac{1}{2}I\omega^2$$

$$\rho = m/V$$

$$\vec{F}_s = k\vec{x}$$

$$U_s = \frac{1}{2}kx^2$$

$$T_s = 2\pi\sqrt{\frac{m}{k}}$$

$$T_p = 2\pi\sqrt{\frac{\ell}{g}}$$

$$T = \frac{1}{f} = \frac{2\pi}{w}$$

$$F_G = \frac{Gm_1 m_2}{r^2}$$

$$U_G = -\frac{Gm_1 m_2}{r}$$

a = acceleration
F = force
f = frequency
h = height
I = rotational inertia
K = kinetic energy
k = spring constant
L = angular momentum
ℓ = length
m = mass
N = normal force
P = power
p = momentum
r = radius or distance
T = period

t = time
U = potential energy
V = volume
v = velocity or speed
ω = angular speed
W = work done on a system
x = position
α = angular acceleration
θ = angle
ρ = density
τ = torque
μ = coefficient of friction

Electricity

$$F = k \cdot \frac{q_1 q_2}{r^2}$$

$$I_{avg} = \frac{\Delta Q}{\Delta t}$$

$$R = \frac{\rho \ell}{A}$$

$$V = IR$$

$$P = IV$$

$$R_s = \sum_i R_i$$

$$\frac{1}{R_p} = \sum_i \frac{1}{R_i}$$

A = area
d = distance
F = force
I = current
ℓ = length
P = power
Q = charge
q = point charge
R = resistance

r = distance
t = time
V = electrical potential
v = velocity or speed
ρ = resistivity
θ = angle

Waves

$$v = f\lambda$$

f = frequency
v = speed
λ = wavelength

Geometry and Trigonometry

Rectangle

$$A = bh$$

Triangle

$$A = \frac{1}{2}bh$$

Circle

$$A = \pi r^2$$

$$C = 2\pi r$$

Parallelepiped

$$V = \ell w h$$

Cylinder

$$V = \pi r^2 \ell$$

$$S = 2\pi r\ell + 2\pi r^2$$

Sphere

$$V = \frac{4}{3}\pi r^3$$

$$S = 4\pi r^3$$

Right Triangle

$$a^2 + b^2 = c^2$$

$$\sin\theta = \frac{a}{c}$$

$$\cos\theta = \frac{b}{c}$$

$$\tan\theta = \frac{a}{b}$$

A = area
C = circumference
V = volume
S = surface area
b = base

h = height
ℓ = length
w = width
r = radius

Glossary

A

acceleration A vector quantity representing the time rate of change of velocity.

action A force applied to an object that leads to an equal but opposite reaction; the product of total energy and time in units of joules times seconds; the product of momentum and position, especially in the case of the Heisenberg uncertainty principle.

alternating current An electric current that changes its direction and magnitude according to a regular frequency.

ammeter A device that, when placed in series, measures the current in an electric circuit; a galvanometer with a low-resistance coil placed across it.

ampere (A) The SI unit of electric current, equal to 1 C/s.

amplitude The maximum displacement of an oscillating particle, medium, or field relative to its rest position.

angstrom (Å) A unit of distance measurement equal to 1×10^{-10} m.

antinodal line A region of maximum displacement in a medium where waves are interacting with each other.

average speed A scalar quantity equal to the ratio of the total distance to the total elapsed time.

B

battery A combination of two or more electric cells.

beats The interference caused by two sets of sound waves with only a slight difference in frequency.

C

Cartesian coordinate system A set of two or three mutually perpendicular reference lines, called axes and usually designated as x, y, and z, that are used to define the location of an object in a frame of reference; a coordinate system named for French scientist René Descartes.

center of curvature A point that is equidistant from all other points on the surface of a spherical mirror; a point equal to twice the focal length of a spherical mirror.

center of mass The weighted mean distribution point where all the mass of an object can be considered to be located; the point at which, if a single force is applied, translational motion will result.

centripetal acceleration The acceleration of mass moving in a circular path directed radially inward toward the center of the circular path.

centripetal force The deflecting force, directed radially inward toward a given point, that causes an object to follow a circular path.

coefficient of friction The ratio of the force of friction to the normal force when one surface is sliding (or attempting to slide) over another surface.

coherent Referring to a set of waves that have the same wavelength, frequency, and phase.

component One of two mutually perpendicular vectors that lie along the principal axes in a coordinate system and can be combined to form a given resultant vector.

concurrent forces Two or more forces that act at the same point and at the same time.

conductor   A substance, usually metallic, that allows the relatively easy flow of electric charges.

conservation of energy   A principle of physics that states that the total energy of an isolated system remains the same during all interactions within the system.

conservation of momentum   A principle of physics that states that, in the absence of any external forces, the total momentum of an isolated system remains the same.

conservative force   A force such that any work done by this force can be recovered without any loss; a force whose work is independent of the path taken.

constructive interference   The additive result of two or more waves interacting with the same phase relationship as they move through a medium.

coordinate system   A set of reference lines, not necessarily perpendicular, used to locate the position of an object within a frame of reference by applying the rules of analytic geometry.

coulomb (C)   The SI unit of electrical charge, defined as the amount of charge 1 A of current contains each second.

Coulomb's law   The electrostatic force between two point charges is directly proportional to the product of the charges and inversely proportional to the square of the distance separating them. Named for French physicist Charles-Augustin de Coulomb.

current   A scalar quantity that measures the amount of charge passing a given point in an electric circuit each second.

cycle   One complete sequence of periodic events or oscillations.

D

damping   The continuous decrease in the amplitude of mechanical oscillations due to a dissipative force.

deflecting force   Any force that acts to change the direction of motion of an object.

derived unit   Any combination of fundamental physical units.

destructive interference   The result produced by the interaction of two or more waves with opposite phase relationships as they move through a medium.

diffraction   The ability of waves to pass around obstacles or squeeze through small openings.

direct current   Electric current that is moving in one direction only around an electric circuit.

displacement   A vector quantity that determines the change in position of an object by measuring the straight-line distance and direction from the starting point to the ending point.

dissipative force   Any force, such as friction, that removes kinetic energy from a moving object; a nonconservative force.

distance   A scalar quantity that measures the total length of the path taken by a moving object.

Doppler effect   The apparent change in the wavelength or frequency of a wave as the source of the wave moves relative to an observer. Named for Austrian physicist Christian Doppler.

dynamics   The branch of mechanics that studies the effects of forces on objects.

E

elastic collision   A collision between two objects in which there is a rebounding and no loss of kinetic energy occurs.

elastic potential energy   The energy stored in a spring when work is done to stretch or compress it.

electrical ground   The passing of charges to or from Earth to establish a potential difference between two points.

electric cell   A chemical device for generating electricity.

electric circuit   A closed conducting loop consisting of a source of potential difference, conducting wires, and other devices that operate on electricity.

electromotive force (emf)   The potential difference caused by the conversion of different forms of energy into electrical energy; the energy per unit charge.

electron   A negatively charged particle that orbits a nucleus in an atom; the fundamental carrier of negative electric charge.

electroscope   A device for detecting the presence of static charges on an object.

elementary charge   The fundamental amount of charge of an electron.

emf   See **electromotive force**.

energy   A scalar quantity representing the capacity to work.

equilibrant   The force equal in magnitude and opposite in direction to the resultant of two or more forces that brings a system into equilibrium.

equilibrium   The balancing of all external forces acting on a mass; the result of a zero vector sum of all forces acting on an object.

escape velocity   The velocity attained by an object such that, if coasting, the object would not be pulled back toward the planet from which it came.

F

field   A region characterized by the presence of a force on a test body like a unit mass in a gravitational field or a unit charge in an electric field.

force   A vector quantity that corresponds to any push or pull due to an interaction of matter that changes the motion of an object.

force constant   See **spring constant**.

forced vibration   A vibration caused by the application of an external force.

frame of reference   A point of view consisting of a coordinate system in which observations are made.

free-body diagram   A diagram that illustrates all of the forces acting on a mass at any given time.

frequency   The number of completed periodic cycles per second in an oscillation or wave motion.

friction   A force that opposes the motion of an object as it slides over another surface.

fundamental unit   An arbitrary scale of measurement assigned to certain physical quantities, such as length, time, mass, and charge, that are considered to be the basis for all other measurements. In the SI system, the fundamental units used in physics are the meter, kilogram, second, ampere, kelvin, and mole.

G

galvanometer   A device used to detect the presence of small electric currents when connected in series in a circuit.

gravitation   The mutual force of attraction between two uncharged masses.

gravity   Another name for gravitation or the gravitational force; the tendency of objects to fall to Earth.

H

hertz (Hz)   The SI unit of frequency, equal to 1 s^{-1}.

Hooke's law   The stress applied to an elastic material is directly proportional to the strain produced. Named for English scientist Robert Hooke.

I

impulse   A vector quantity equal to the product of the average force applied to a mass and the time interval in which the force acts; the area under a force versus time graph.

inelastic collision   A collision in which two masses interact and stick together, leading to an apparent loss of kinetic energy.

inertia   The property of matter that resists the action of applied force trying to change the motion of an object.

inertial frame of reference   A frame of reference in which the law of inertia holds; a frame of reference moving with constant velocity relative to Earth.

instantaneous velocity   The slope of a tangent line to a point in a displacement versus time graph.

insulator   A substance that is a poor conductor of electricity because of the absence of free electrons.

interference   The interaction of two or more waves, producing an enhanced or a diminished amplitude at the point of interaction; the superposition of one wave on another.

interference pattern   The pattern produced by the constructive and destructive interference of waves generated by two point sources.

isolated system   A combination of two or more interacting objects that are not being acted upon by external force.

J

joule (J)   The SI unit of work, equal to $1 \text{ N} \cdot \text{m}$; the SI unit of mechanical energy, equal to $1 \text{ kg} \cdot \text{m}^2/\text{s}^2$.

junction   The point in an electric circuit where a parallel connection branches off.

K

Kepler's first law   The orbital paths of all planets are elliptical. Named for German astronomer Johannes Kepler.

Kepler's second law   A line from the Sun to a planet sweeps out equal areas in equal time.

Kepler's third law   The ratio of the cube of the mean radius to the Sun to the square of the period is a constant for all planets orbiting the Sun.

kilogram (kg)   The SI unit of mass.

kilojoule (kJ)   A unit representing 1,000 J.

kinematics   In mechanics, the study of how objects move.

kinetic energy   The energy possessed by a mass because of its motion relative to a frame of reference.

kinetic friction   The friction induced by sliding one surface over another.

Kirchhoff's junction rule   The algebraic sum of all currents at a junction in a circuit equals zero. Named for German physicist Gustav Kirchhoff.

Kirchhoff's loop rule   The algebraic sum of all potential drops around any closed loop in a circuit equals zero.

L

law of inertia   See **Newton's first law of motion**.

longitudinal wave   A wave in which the oscillating particles vibrate in a direction parallel to the direction of propagation.

M

mass   The property of matter used to represent the inertia of an object; the ratio of the net force applied to an object and its subsequent acceleration as specified by Newton's second law of motion.

mean radius   For a planet, the average distance to the Sun.

meter (m)   The SI unit of length.

moment arm distance   A line drawn from a pivot point.

momentum   A vector quantity equal to the product of the mass and the velocity of a moving object.

N

natural frequency   The frequency with which an elastic body will vibrate if disturbed.

net force   The resultant force acting on a mass.

newton (N)   The SI unit of force, equal to $1 \text{ kg} \cdot \text{m}/\text{s}^2$.

Newton's first law of motion An object at rest tends to remain at rest, and an object in motion tends to remain in motion at a constant velocity, unless acted upon by an external force; the law of inertia. Named for British mathematician/physicist Sir Isaac Newton.

Newton's second law of motion The acceleration of a body is directly proportional to the applied net force; $\vec{F} = m\vec{a}$.

Newton's third law of motion For every action, there is an equal, but opposite, reaction.

nodal line A line of minimum displacement when two or more waves interfere.

nonconservative force A force, such as friction, that decreases the amount of kinetic energy after work is done by or against the force.

normal A line perpendicular to a surface.

normal force A force that is directed perpendicularly to a surface when two objects are in contact.

O

ohm (Ω) The SI unit of electrical resistance, equal to 1 V/A.

Ohm's law In a circuit at constant temperature, the ratio of the potential difference to the current is a constant (called the resistance). Named for German physicist Georg Ohm.

P

parallel circuit A circuit in which two or more devices are connected across the same potential difference and provide an alternative path for charge flow.

period The time, in seconds, to complete one cycle of repetitive oscillations or uniform circular motion.

phase The relative position of a point on a wave with respect to another point on the same wave.

polar coordinate system A coordinate system in which the location of a point is determined by a vector from the origin of a Cartesian coordinate system and the acute angle it makes with the positive horizontal axis.

potential difference The difference in work per unit charge between any two points in an electric field.

potential energy The energy in a system resulting from the relative position of two objects interacting via a conservative force.

power The ratio of the work done to the time needed to complete the work.

pressure The force per unit area.

pulse A single vibratory disturbance in an elastic medium.

R

resistance The ratio of the potential difference across a conductor to the current in the conductor.

resistivity A property of matter that measures the ability of that substance to act as a resistor.

resolution of forces The process by which a given force is decomposed into a pair of perpendicular forces.

resonance The production of sympathetic vibrations in a body, at its natural vibrating frequency, caused by another vibrating body.

rotational equilibrium The situation when the vector sum of all torques acting on a rotating mass equals zero.

rotational inertia The ability of a substance to resist the action of a torque; moment of inertia.

S

scalar A physical quantity, such as mass or speed, that is characterized by magnitude only.

second (s) The SI unit of time.

series circuit   A circuit in which electrical devices are connected sequentially in a single conducting loop, allowing only one path for charge flow.

sliding friction   A force that resists the sliding motion of one surface over another; kinetic friction.

speed   A scalar quantity measuring the time rate of change of distance.

spring constant   The ratio of the applied force and resultant displacement of a spring.

standing wave   A stationary wave pattern formed in a medium when two sets of waves with equal wavelength and amplitude pass through each other, usually after a reflection.

static electricity   Stationary electric charges.

static friction   The force that prevents one object from sliding over another.

statics   The study of the forces acting on an object that is at rest relative to a frame of reference.

superposition   The ability of waves to pass through each other, interfere, and then continue on their way unimpeded.

T

torque   The application, from a pivot, of a force at right angles to a designated line that tends to produce circular motion; the product of a force and a perpendicular moment arm distance.

total mechanical energy   In a mechanical system, the sum of the kinetic and potential energies.

transverse wave   A wave in which the vibrations of the medium or field are at right angles to the direction of propagation.

U

uniform circular motion   Motion around a circle at a constant speed.

uniform motion   Motion at a constant speed in a straight line.

unit   An arbitrary scale assigned to a physical quantity for measurement and comparison.

universal law of gravitation   The gravitational force between any two masses is directly proportional to the product of their masses, and inversely proportional to the square of the distance between them.

V

vector   A physical quantity that is characterized by both magnitude and direction; a directed arrow drawn in a Cartesian coordinate system used to represent a quantity such as force, velocity, or displacement.

velocity   A vector quantity representing the time rate of change of displacement.

volt (V)   The SI unit of potential difference, equal to 1 J/C.

voltmeter   A device used to measure the potential difference between two points in a circuit when connected in parallel; a galvanometer with a high-resistance coil placed in series with it.

W

watt (W)   The SI unit of power, equal to 1 J/s.

wave   A series of periodic disturbances in an elastic medium or field.

wavelength   The distance between any two successive points in phase on a wave.

weight   The force of gravity exerted on a mass at the surface of a planet.

work   A scalar measure of the relative amount of change in mechanical energy; the product of the magnitude of the displacement of an object and the component of applied force in the same direction as the displacement; the area under a graph of force versus displacement.

Index